W9-AEO-895

ARTIFICIAL VISION FOR ROBOTS

VOL 1

ARTIFICIAL VISION FOR ROBOTS

Edited by
Professor I Aleksander

ASQC
LIBRARY

CHAPMAN & HALL
NEW YORK

Chapters 2 to 10 inclusive were first published in *Digital Systems for Industrial Automation* by Crane Russak & Company Inc, 3 East 44th Street, New York, NY 10017, USA.

This collected edition first published in Great Britain in 1983 by Kogan Page Ltd, 120 Pentonville Road, London N1 9JN; first published in the United States of America in 1984 by Chapman & Hall, 733 Third Avenue, New York, NY 10017.

Copyright © Chapters 2 to 10 inclusive Crane Russak & Company Inc 1981 and 1982
Copyright © Chapters 1, 11 and 12 Igor Aleksander 1983

All rights reserved. No part of this book may be reprinted, or reproduced or utilized in any form or by any electronic, mechanical or other means, now known or hereafter invented, including photocopying and recording, or in any information storage or retrieval system, without permission in writing from the publishers.

ISBN: 412 004518

Printed and bound in Great Britain

Contents

Chapter 1

Introduction

I. ALEKSANDER

Department of Electrical Engineering and Electronics
Brunel University, England

The three key words that appear in the title of this book need some clarification.

First, how far does the word *robot* reach in the context of industrial automation? There is an argument maintaining that this range is not fixed, but increases with advancing technology. The most limited definition of the robot is also the earliest. The history is worth following because it provides a convincing backdrop to the central point of this book: vision is likely to epitomize the technological advance, having the greatest effect in enlarging the definition and range of activity of robots.

In the mid 1950s it was foreseen that a purely mechanical arm-like device could be used to move objects between two fixed locations. This was seen to be cost-effective only if the task was to remain fixed for some time. The need to change tasks and therefore the level of programmability of the robot was a key issue in the broadening of robot activities. Robots installed in industry in the early 1960s derived their programmability from a device called a *pinboard*. Vertical wires were energized sequentially in time, while horizontal wires, when energized, would trigger off elementary actions in the manipulator arm. The task of reprogramming was a huge one, as pins had to be reinserted in the board, connecting steps in time with robot actions. Thousands of pins might have been involved, requiring many hours, if not days, of work.

In the 1960s programmability became available through connections with large mainframe computers. Despite the awkwardness of the umbilical connections that such a methodology demanded, large motor car producers were not deterred from developing plans for the deployment of armies of manipulator arms in the welding of cars on automatic production lines. The automatic Fiat car assembly plant in Turin was the product of such thinking.

In the meantime the initiative taken by the Digital Equipment Corporation in putting on the market small machines, brought the next step in robot programmability. Contrary to the expectations of the manufacturers of large mainframe computers (for instance IBM), the computing power offered by these minicomputers was more cost-effective than that available through connections with large mainframe computers. The reason is simple. The price of electronic parts of a computer was plummeting at the same time as the reliability of the manufacture of semiconductor devices was going up. Manpower costs became the central item in the overheads of computer manufacture. In this way, by stepping down computer size, economies of scale favoured the small-but-many rather than the large-but-few strategy. This was certainly true for robot manufacturers who could sell fully controlled robots with their own individual computer, without having to rely on the presence of large mainframe systems before being able to achieve a sensible level of programmability.

The advent of the microprocessor has taken this development several stages further. Control boxes for robots have, for some years, contained several fully fledged computers in microprocessor form. This ensures that not only can sophisticated programs be stored, but also that several functions (such as control strategies) may be computed independently, at high speed. This large amount of computing power can be contained in a space not much bigger than a shoebox.

Despite the increase in programmability, the robot was still unable to depart from a single predetermined sequence of actions. Such departures would be necessary should the robot be required to take different actions appropriate to variations in events in its operating space. A very simple example of such 'branching' is a program that instructs the robot to 'put it to the left if it is round, or to the right if it is square'. In order to execute such branching the robot would, in some way, have to *sense* whether the object was round or square. In the opinion of most of the contributors to this book, vision for robots is the most powerful available mode of sensing such events. This would not only produce a massive step forward in the range of applicability of robots, but also have a revolutionary effect on the way that robots are used. It might also provide more 'intelligence' for other automatic manufacturing systems. The commonality of techniques of vision employed across these fields would spread the ethos of robotics to all forms of automation.

In 1980, Joseph Engelberger, the first industrial entrepreneur to take robots from the laboratory into the factory, whose activities thrive on the sales of sightless robots wrote[1];

> 'To date, robots have largely been insensate, but roboticists are striving to correct this deficiency.When robots do boast of sight and touch the list of applications . . . will merit a large supplement;'

Indeed, the plans of Engelberger's Unimation Inc are known to be well advanced in the area of vision.

Some examples may serve to underpin the expectations being vested in robot vision. To make a general point, it may be worth thinking of the handicap suffered by a human being deprived of sight. It is known that in order to overcome some of the limitations a sightless person develops alternative skills. For example, the blind person may develop far greater control over his own muscles (ie, increased sensitivity) to restore some of the precision he needs to position his limbs, having lost this ability through blindness. Not only is his control over his muscles improved, but also he becomes more sensitive to the feedback which the nerve endings in his muscles are providing. In sightless robot technology, much of the precision that has been engineered into current manipulator arms has, in a way, been necessary to compensate for the lack of sight.

Vision will reverse the process. It will provide sensory feedback instead of relying on open-loop control. As will be seen in many of the papers in this book, it is argued that vision is now an economic possibility. Its cost-effectiveness is derived from microchip technology, which is decreasing in price with time. This serves to offset the increasing costs that would be demanded from mounting mechanical precision sought in sightless installations.

An example may serve to illustrate the way that the role of a robot might be redefined in an industrial setting. Imagine that a sightless robot is used to move objects arriving via conveyor belt A to either conveyor belt B or conveyor belt C where they will be carefully channelled and marked by a paintbrush placed in a fixed position. These objects might be bonnets and tailgates of toy cars which eventually would be painted blue and green respectively. Consider the technical problems. First, conveyor belt A would have to contain jigs that present the parts to the manipulator in a predetermined manner. The manipulator controller would have to be told exactly how many parts of each kind were contained in each batch, so that the control program would know when to branch between one part of the program and the other. Now imagine the presence of a vision camera and a recognition system that can be pointed at conveyor belt A. Clearly there would be no need for special jigs as the vision system could not only identify the parts but also inform the manipulator as to where they were placed. Then, armed with a paintbrush, the robot could mark the parts appropriately without having to pick them up. This illustrates the way in which the use of vision transfers the development overheads from mechanical complexity and precision to visual processing which, as already stressed, benefits from favourable microprocessing economics.

The need for good vision for robots thus becomes undeniable, but is it technologically feasible?

The Front End

The vidicon camera is the obvious vision transducer for most general purpose robotic tasks. An industrial standard is developing with respect to the choice of sampling intervals in the sense that a square picture of 512×512 dots is usually extracted. Such dots are called *pixels* and are rarely composed of more than 8 bits. That is, each dot can be sensed in one of 256 (ie 2^8) levels of greyness. In many practical cases images of 64×64 pixels and 16 (ie 2^4) grey levels are adequate and provide a very inexpensive transducer.

Equally important as vidicon devices are photodiode solid-state cameras or charge-coupled device (CCD) cameras. It would be inappropriate to give a detailed technical account of these transducers here. Suffice it to say that their feasibility is well-established. For a detailed account of such devices refer to *Robot Technology Volume 2: Interaction with the Environment* by Coiffet [2].

Almost as important as the transducer itself, is the memory that holds the image while the vision processor operates on the image. These devices are called *frame stores*, and are currently being developed as standardized items of equipment.

Again, silicon-integrated technology has played a great part in increasing the cost-effectiveness of this part of the set of devices that make up a vision system. Not only are such devices capable of storing single frames, but also they include generally a number of useful processing functions. An important example of this is found in the possibility of extracting information from a window of selectable size from the full image. Also, some simple image processing tasks can be executed in the frame store itself. At this stage it may be worth explaining some of the terms used in robot vision studies.

Image Processing

Image processing has to be distinguished from *pattern recognition*. Image processing implies a transformation of an image, whilst pattern recognition makes a statement regarding the content of the image.

Much can be achieved on the strength of image processing alone. For example, some objects could be distinguished solely by measurement of parameters such as area and perimeter of a silhouette-type shape. Both of these tasks would fall under the heading *image processing*.

Another typical image processing task is noise removal. Again, this type of processing can be carried out by the frame store using the following technique. A very small window (ie 3 ×3 pixels) scans the exterior of the entire image. A simple detection mask can be made to remove any parts of the image that appear isolated.

Vision systems vary much in terms of how much reliance is placed on image processing, and how much of a given method could be described as pattern recognition. Examples of systems stressing each of these two methodologies will be found in this book. Usually the more general methods rely more heavily on pattern recognition, whereas the more problem-oriented systems tend to use image processing techniques.

Pattern Recognition

The history of pattern recognition dates back to the mid 1950s and is rich with work ranging from the highly practical to the obstrusely theoretical. The technical details concerning the major trends of this branch of computer science or engineering are left for the main body of this book (see Aleksander, Stonham and Wilkie, Chapter 10).

Here, it may be worth qualifying some of the issues that make pattern recognition a central necessity for most robot systems. In pure form, a pattern recognizer receives an image as input, and produces a statement as to which of a limited number of objects the image belongs. The two approaches employed are *adaptive* and *algorith-*

mic. The latter is employed when the objects to be recognized are well-defined. A typical application of algorithmic techniques is the recognition of codes printed in a style designed for the purpose. These are familiar objects on most bank cheques, and have become well known as a symbol for computerization. What if non-stylized characters, typewritten or even hand written characters could be recognized? It was precisely a set of such questions that led to many developments of the first kind. With this there is usually associated a learning phase during which examples of variants of each character are shown to the system.

As a result of these examples the adaptive system calculates the key differences between different objects and the similarities of objects within a particular class. This learning approach gives a system much flexibility and does not limit the performance of the system to predetermined shapes defined by a programmer. There are many vision systems on the market that lie between the algorithmic and the adaptive modes. These operate on the basis of extracting features from images in a pre-programmed way, and attach classification labels to groups of these features, thus providing the learning element. For example, the pre-programmed feature extraction process may consist of measuring the area of an object as feature 1, its perimeter as feature 2, its major radius from the centre of area as feature 3 etc. Then, if object A is presented to the system, its set of features is labelled A, and so on for other features. The difficulty with this approach is that it requires long and time-consuming searches through the feature data. The last section of this book will deal with hardware methods that avoid these serious time delays. However, systems that are designed to maximize the use of standard as opposed to special architectures are of great importance in obtaining rapid solutions to immediate vision problems in contemporary robots.

Applications

One way of discussing the applications of robot vision is to list those things that clearly cannot be achieved by the sightless robot. This, however, may be the wrong way to approach it. If it is assumed that a robot has vision as a matter of course, it is soon discovered that the robot is better able to carry out tasks that even sightless robots can do, and does so more efficiently.

'Pick-and-place' is the name given to simple transportation tasks, from, say, one work surface to another. In sightless manipulators, as already described, the workpieces have to arrive at a well-defined position in the task space. Vision removes this constraint and hence a robot might be able to select appropriate parts from a heap of workpieces in the task space. Some of the ways of doing this are included in the paper by Hättich (see Chapter 5).

Having identified the workpieces, the next problem for a robot is to decide on a way of moving the gripper in order to pick up the required part. This problem is discussed, together with some ways of solving it, in the paper by Page and Pugh (see Chapter 7) where methods of choosing the appropriate gripper and then deploying it are presented.

A commonly encountered application of vision in robotics is the possibility of using visual feedback in assembly tasks. Practice in sightless robots has suggested that assembly can be achieved only under conditions where designers have a special degree of control over the shape of the workpieces. For example, to fit a dowel into a hole, chamfers would have to be machined to both parts of a two-part assembly, and, in more complex tasks, special end stops or guide slots would have to be machined to allow for the lack of 'knowledge', in the controlling computer, of the *precise* location of the end effector and the piecepart it holds. The paper by Saraga and Jones (see Chapter 6) describes the use of vision in avoiding the need for special piecepart designs.

Many applications depend on the results of about 20 years of work in artificial intelligence (AI) laboratories throughout the world. This concerns endowing the robot with the ability to solve problems and develop strategic plans. Traditionally this was seen as a necessity for mobile robots to find their way about in an unknown environment, as might occur in space exploration. Unfortunately, conventional computing methods have proved to be too slow and too demanding of the computer memory even in highly simplified situations. The last paper in this book (see Aleksander, Chapter 12) presents an approach to the reasoning power, based on special architectures, that has grown out of the need to process vision data very fast.

Finally, the three applications generally considered by the robot vision research community to be exceptionally difficult should not be forgotten. These are bin picking, depalletization and quality control. Bin picking, as the name implies, is the ability to pick up a part from a jumble of equal parts lying heaped up in a bin. Depalletization is a similar problem, but one in which the parts are arranged on a pallet in a more ordered way. Quality control demands the separation of faulty or damaged parts from good ones, even if the nature of the damage is very slight. These applications have one characteristic in common, and that is the likelihood that the parts cannot be seen by the vision system as silhouettes. It is for this reason that Part III is devoted to a methodology aimed at solving such non-silhouette problems, even though it has, as yet, not become industrial practice.

Structure of the Book

This book consists largely of selected published papers in the field of robot vision. They represent the most recent developments not only in academic institutions, but also in industrial research laboratories. The other eleven papers have been grouped into three main parts:

I. Techniques;
II. Applications;
III. Adaptive processing for vision.

I. Techniques: The science of robot vision draws from a wide background of techniques ranging from computer science through mathematics to electronics. For the newcomer to the subject, the selection of appropriate methods to solve a particular problem in robot vision is not an obvious process.

The first selection is the decision to implement a given algorithm with hardware, or software, or a mixture of the two. In particular, this occurs at the image processing level where the construction of a special purpose image store must be considered as an alternative to classical programming for a minicomputer or a microcomputer. The paper by Dessimoz and Kammenos (see Chapter 2) is a well-developed guide on this subject, based on the authors' experience. Their presentation centres on effective ways of processing images of back-lit silhouettes of complex industrial pieceparts. The algorithms investigated concern the following techniques. First, *image filtering* is discussed in which the algorithm removes jagged edges that can arise either through the necessity of fitting angled straight lines on rectangular image grids or electrical noise in the imaging camera. *Skeletonization* is the second technique presented. This is the process of representing a solid object by a skeleton which may be sufficient either to identify the object or its orientation. *Curve smoothing* and *polar mapping* are also discussed in some detail. The first of these methods is similar to image filtering, but is applied to curves rather than to straight lines. The second is a very widely used technique present in commercially available vision systems for robots. It consists of finding the centre of the object and then measuring the distance of important features (eg corners) from it. The paper ends by describing some methods of representing an object by means of its features and the way in which they relate to one another. The authors' general conclusion centres on the idea that good system design for vision will, in the future, be more a case of structuring a system from standard image processing modules made in hardware, rather than the programming of a general purpose computer.

A similar survey, but on the topic of *recognition* is pursued by Pot, Coiffet and Rives (see Chapter 3). They have sifted through a long history of work in pattern recognition and distilled five methods relevant to robot scene analysis. The task chosen as a vehicle for their comparative study is the identification of a set of solid objects under well-lit conditions. An image processing scheme similar to that described in Chapter 2 is used to extract polar coordinate features from these objects. The aim of the paper is to describe the efficacy of the five chosen methods in identifying the originating object from the feature sets. The presentation of an object always results in a list of numbers each of which represents the degree of 'presence' of each feature. All five methods make use of the overall difference between the feature list generated by an unknown object and that known about such lists from a training period. In practice, the scheme should be used with an experimental system that captures an image on a 100×100 photodiode camera. The authors' conclusion centres on the fact that a particular statistical method of storing the result of the training sequence yields optimal results.

The last contribution in this section, by Gaglio, Morasso and Tagliasco (see Chapter 4) creates a bridge between pattern recognition and artificial intelligence. Again, both image processing and pattern recognition are involved in a real-data system in which the image is captured by a TV camera and a 512×512, 8-bit pixel camera and digitizer. An edge-detection technique is used for initial image processing to which a language-like processing system is applied. This, blended with a statistical storage process, is used in a further linguistic process not only to describe objects, but also to describe scenes. The authors find an interesting parallel in the process used to describe the human perceptual system [3].

II. Applications: The common characteristic of the papers in this section is that they deal with realistic situations found in the production industry. The first paper by Hättich (see Chapter 5) concentrates on scenes that contain engineering parts such as metal flanges that include bends in the metal and industrial bolts. In a manner somewhat reminiscent of the last paper discussed, the scheme relies on internal models of the workpieces, these being built up during training. This time, however, the inner models are neither linguistic nor polar coordinate sets, but sets of line segments and the relations between them. An interesting method of storing such models as state transition diagrams is worthy of note for wider applications. Although the algorithms and tests presented only deal with straight lines, remarkably good results are obtained even with heavily overlapping objects. The author is optimistic that this method can be extended to mixtures of lines and curves.

The second paper in this section, by Saraga and Jones (see Chapter 6), is unique as it demonstrates the way in which a system of multiple cameras can be coordinated in a high-accuracy assembly task. Of particular interest is the use of parallax between two cameras. An assembly accuracy of one-tenth of a millimetre was achieved using 600×600 pixel cameras.

This is followed by the work of Page and Pugh (see Chapter 7) which tackles the end effector orientation problem, the solution of which is a prerequisite to successful bin picking or depalletization. A central criterion that the authors set themselves is to employ a conventional computer and to reduce the computational overheads to a minimum. The processing times achieved on 128×128 pixel images were between 1 and 4 seconds which, as the authors stress, are generally faster than those of a human operator. The system also demonstrably resides in 6 kilowords of 16-bit-per-word memory. This paper is typical of the philosophy developed at the Department of Electronic Engineering of the University of Hull that simplicity in program structure helps to resolve many practical problems.

The last paper in this sequence (see Chapter 8) is short, but highly practical. It demonstrates the simplicity with which the front end of a vision system can be designed. The paper by Todd is purely a statement on electronic design. It is convincing in that it shows how a simple integrated circuit array may be interfaced to a rudimentary microprocessor chip, and for less than the cost of a motor car battery an imaging device for a robot can be constructed. In this specific application, the system was used as a range-finder for a mobile robot.

III. Adaptive processing for vision: This section of the book contains four papers describing one particular development at Brunel University in England. It is an in-depth study of the WISARD system (derived mnemonically from the names of the authors: WIlkie, Stonham and Aleksander's Recognition Device). This is an adaptive, general purpose recognition device, which is *taught* its task by an operator. The WISARD philosophy is somewhat different from the pattern recognition strategies discussed in the rest of the book. It is a special network of microcircuit memory units operating with a frame store that derives an image from a television camera. The system can equally well receive input from almost any imaging device. It shows examples of images together with their desired classifications, and subsequently classifies unknown images into the taught classes. Being, in theory, a totally parallel system, it can provide up to two million decisions per second. In practice, such speeds are excessive, particularly since imaging devices can currently deliver only 50 images per second for standard TV cameras, increasing to a few hundred images for solid-state cameras. It is for this reason that the WISARD system is organized in a semi-parallel way so that it operates faster than conventional computing systems without the expense of total parallelism.

The first paper by Stonham (see Chapter 9) discusses the fundamental operation of memory networks, showing the way in which their performance depends on the size of the network and other important parameters. Some applications, both within and outside robotics, are discussed to demonstrate the generality of the method.

The second paper (see Chapter 10) compares the WISARD methodology to some of the other standard approaches, some of which have been described in the earlier parts of this book. It is shown that given special purpose construction of the computing equipment, the parallel or semi-parallel systems as employed in the WISARD system have a higher potential for cost-effective performance than other special purpose schemes. Also, some of the design decisions made during the design of the WISARD systems are discussed. It considers more deeply the applications of a good, general purpose pattern recognizer. This supports the point made earlier that good vision can extend the definition of the robot itself. The paper emphasizes that an artificial system that recognizes, for example, false signatures, intruders in a secure environment or identifies diseases from symptoms presented, is as necessary to the appellative robot as a device with hands, wrists and elbows.

The third paper (see Chapter 11) in the sequence is aimed at the designer who wishes to use the WISARD system specifically in one of the robotic applications discussed earlier. There is a description of the theory of the system in simplified form, usable by a designer. The paper bases its design calculations on the ability of the system to be sensitive to changes of shape of objects. Questions of recognizing objects appearing in any position within a field of view, and how network parameters to do this are chosen are answered. Also the way in which these systems interpolate between items of taught data is discussed. In particular, this interpolative property is useful in measuring the orientation of the part. It is shown that the method applies to overlapping parts and provides an alternative to the methodology offered by Hättich (see Chapter 5) in Part II.

The section ends (see Chapter 12) with the most recent work in this area, which shows that memory networks can become the components of other structures with more intelligent properties. Although much of this paper is speculative, it lists the nature of such intelligent properties and their possible applications. For example, feedback of the net's decision to the input together with the image being recognized are shown to have a major effect on the confidence with which the system can distinguish even minor differences in the input image, should such a distinction be indicated by the user during the training procedure. Feedback is in fact the key to many of the advanced properties of such systems. A second set of properties of such systems is their ability to sustain an inner image (a mental image?) of the objects in view. This enables the robot to take actions with this inner image as a reference. This is of potential use in assembly tasks.

To complete this paper, a third form of feedback is suggested, analogous to that which exists between the eye, the brain and the eye muscles in natural vision. This gives the robot eye the ability to 'look around' objects, a helpful adjunct both in object recognition and assembly strategy planning.

Future Directions for Robot Vision

It is almost certain that vision will become standard in robots. Almost certainly, such systems will contain elements of all the approaches represented in this book. It may therefore be constructive to relate the likely standard features of a future system to some of these ideas. This model system would, obviously, have one or more input cameras, easily interfaced to a micro-controlled temporary store (frame store) as described by Todd (see Chapter 8). The frame store itself will have built-in intelligence which means that by selecting an appropriate mix of instructions, both recognition and image processing algorithms surveyed in Part I should be deployed. The system will also contain a battery of conventional microprocessors operating in a coordinated way, equipped with gripper control algorithms such as discussed by Page and Pugh (see Chapter 7). The robots will be capable of making decisions for assembly tasks based on multiple camera views – and in-camera images of the end effector. Such algorithms will be helped by the data derived from experiments such as described by Saraga and Jones (see Chapter 6). Finally, besides having standard memory, the system will be equipped with an adaptive memory of the WISARD kind which will remove the responsibility for programming from the robot operator. All that is required of the operators is to *teach* the machine the final subtleties that the desired task demands.

References

1. Engelberger, J. F. *Robotics in Practice* Kogan Page, 1980.
2. Coiffet, P. *Robot Technology Volume 2: Interaction with the Environment* Kogan Page, 1983.
3. Linsay, P. H. and Norman, D. A. *Human Information Processing: An Introduction to Psychology* Academic Press, New York, 1972.

Part I: **Techniques**

Chapter 2

Software or Hardware for Robot Vision?

J. D. DESSIMOS* and P. KAMMENOS

Laboratoire de traitement des signaux
Ecole Polytechnique Fédérale de Lausanne
Switzerland

Abstract The optimality of algorithms for the same task may differ depending on the host equipment: specialized hardware or general-purpose computer.

Several methods developed for scene analysis in robotics are presented as examples of strategies specific either to hardwired logic or to a programmed processor.

While analytic, sequential, computed, and structural methods tend to be more suitable for implementation on general-purpose computers, corresponding iterative, parallel, tabulated, and correlative solutions are more easily implemented on logic boards.

Image filtering, contour skeletonization, curve smoothing, polar mapping, and structural description for orienting industrial parts are typical tasks illustrating how the host equipment influences the choice of algorithms.

Introduction

Industrial automation is a field of growing interest driven by the need for breakthroughs in productivity. Many developments are being reported where vision is included in the control loop that drives production tools, e.g., [1, 2]. The Signal Processing Laboratory of the Swiss Federal Institute of Technology in Lausanne has been studying a robot vision system for the past few years. Several results on 2-D scene analysis have been reported [3–12]. It is well known that for the same task, different algorithms may be optimal (in speed, accuracy, and cost) depending on the host equipment. Specialized hardware is always faster than a general-purpose computer (GPC). However, such a computer is best suited to simple algorithms.

The question usually arises very early for a given application whether

*J. D. Dessimoz is now with the University of Rhode Island, USA.

to develop a software program on a general-purpose computer, or to design a specialized hardwired processor. This paper presents several methods developed for scene analysis in robotics, as examples of strategies more specific to hardwired logic or to a programmed processor. Image filtering, skeletonization and smoothing of contours, polar mapping, and structural description for orienting industrial parts are examples of typical strategies.

Figure 1. Original picture

Direct versus Iterative Implementation

Iteration is a way to achieve highly complex solutions through the repeated use of a simple structure. For instance, while a computer can directly convolve a wide Finite Impulse Response (FIR) filter with a picture, specialized processors can obtain a similar result by applying a local operator several times. Although only a small class of wide-window operators can be rigorously synthetized by the iterative use of a small window, most such operators can be satisfactorily approximated in that way.

0.000	0.000	0.001	0.002	0.001	0.000	0.000
0.000	0.003	0.013	0.022	0.013	0.003	0.000
0.001	0.013	0.059	0.097	0.059	0.013	0.001
0.002	0.022	0.097	0.159	0.097	0.022	0.002
0.001	0.013	0.059	0.097	0.059	0.013	0.001
0.000	0.003	0.013	0.022	0.013	0.003	0.000
0.000	0.000	0.001	0.002	0.001	0.000	0.000

0.111	0.111	0.111
0.111	0.111	0.111
0.111	0.111	0.111

Figure 2. A gaussian filter (a) applied to the picture of Figure 1 leads to smooth results (b). The iterative use (3 iterations) of a simple operator (c) applied to the same picture leads to similar results (d)

Let us examine a lowpass filter in more detail. Sharp transitions of the impulse response lead to poor out-of-band rejection. Therefore, a gaussian-shaped window is often recommended.

A filter carries out the convolution of a "window" (point-spread function) h(k, l) with a picture b(k, l). The filtered picture is expressed as follows:

$$b(k,l) = \sum_{m}\sum_{n} b(m.n) \times h(k-m, l-n) \qquad (2\text{-}1)$$

where the window is of size m × n.

The simplest way of carrying out the convolution on a GPC consists in computing explicitly equation (2-1) (Fig. 2a, b).

A specialized processor, however, can use a simple square window ("averager") iteratively. The necessary structure is then much simpler, and does not even require any multiplication [8]. Yet the process tends toward a gaussian filtering because a large number of convolutions leads to a gaussian window (Fig. 2c, d).

Sequential versus Parallel Algorithms

Parallel operation is fast but sequential processing may be more selectively applied.

Consider, for example, the task of skeletonizing thick contours (Fig. 3). Two fundamentally different techniques are particularly suitable. One consists in processing the whole image with small logical operators as suggested by Blum [13] and Arcelli [14] (Fig. 4). The other consists in skeletonizing contour segments one at a time, quasi-simultaneously, with a semantic interpretation of the temporary results [10] (Fig. 5).

The former method leads to homogeneous results throughout the picture (Fig. 4c), and is well suited, due to its simplicity, to (parallel) specialized circuits. If a picture is represented as a set of little square tiles on a cartesian grid, 4-connectivity characterizes picture elements that have a common face, while 8-connectivity also includes the elements with a common corner. An n-connected chain is a sequence of n-connected picture elements (pixels). Consider a contour element P and its 8 closest neighbors in the image (Fig. 4a, b). "Zero" denotes the background. The removal condition can basically be expressed as follows:

> If there is exactly one n-connected chain that contains all the non-zero neighbors of P, then P can be removed.

ASQC
LIBRARY

Figure 3. Thick contours

With n-connectivity in the removal condition, an n-connected skeleton is obtained.

The latter strategy is preferable for implementation on a GPC and also when high-level contextual information is required. For instance, erosion can be selectively inhibited in the vicinity of gaps in the direction of the previously extracted skeleton (Fig. 5).

Consider a small logical operator tracking the edge of a thick contour. Two following constraints should be satisfied:

The step towards the next edge element should be of minimal size in order to avoid leaving the edge and jumping to another contour.

The angle defined by three consecutive tracker positions, past, present, and future, is always maximal (minimal); this insures that the tracker does not enter the contour, but keeps on tracking its "left" ("right") edge.

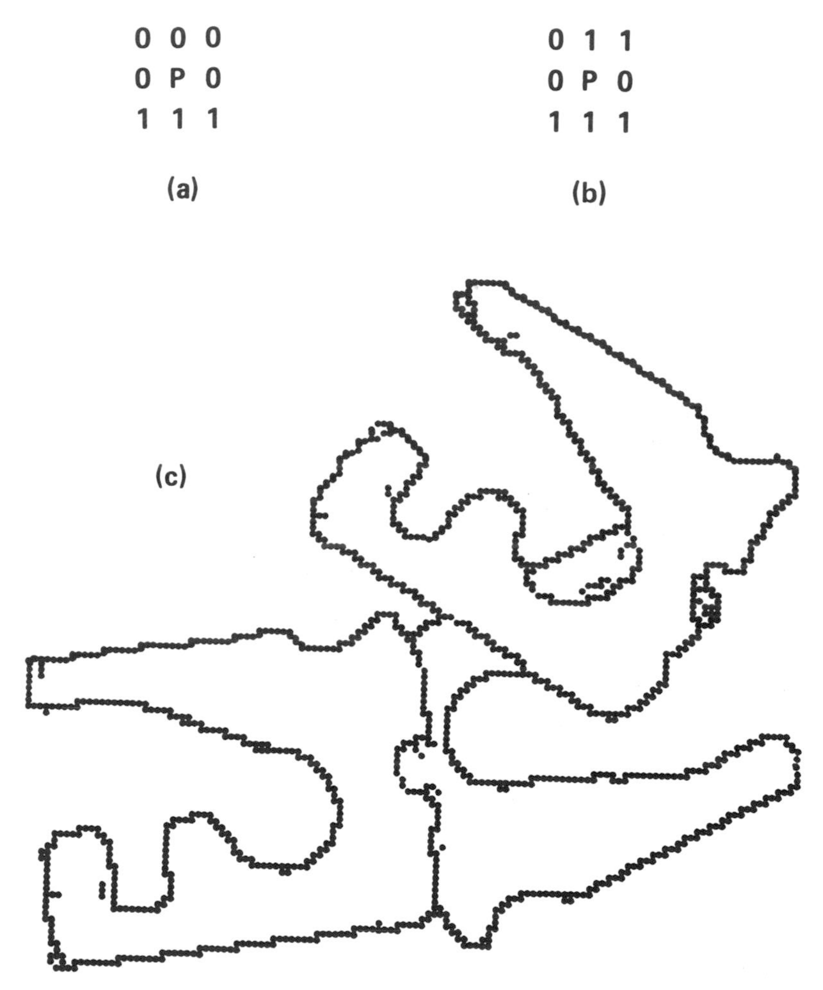

Figure 4. Parallel line-thinning. Examples of contour point configurations where the central pixel can be removed (a) or must be kept (b). Contours of Figure 3 after thinning (c)

Fig. 5a illustrates the priorities of 19 neighbors of an edge tracker when choosing the next tracker position.

When two complementary trackers (''left'' and ''right'') are located on each side of a thick line, and are synchronized by a common third constraint of minimum distance between them, the line skeleton is easily estimated as the locus of their average position at each step (Fig. 5). Moreover, line crossings are detected when the trackers diverge, and

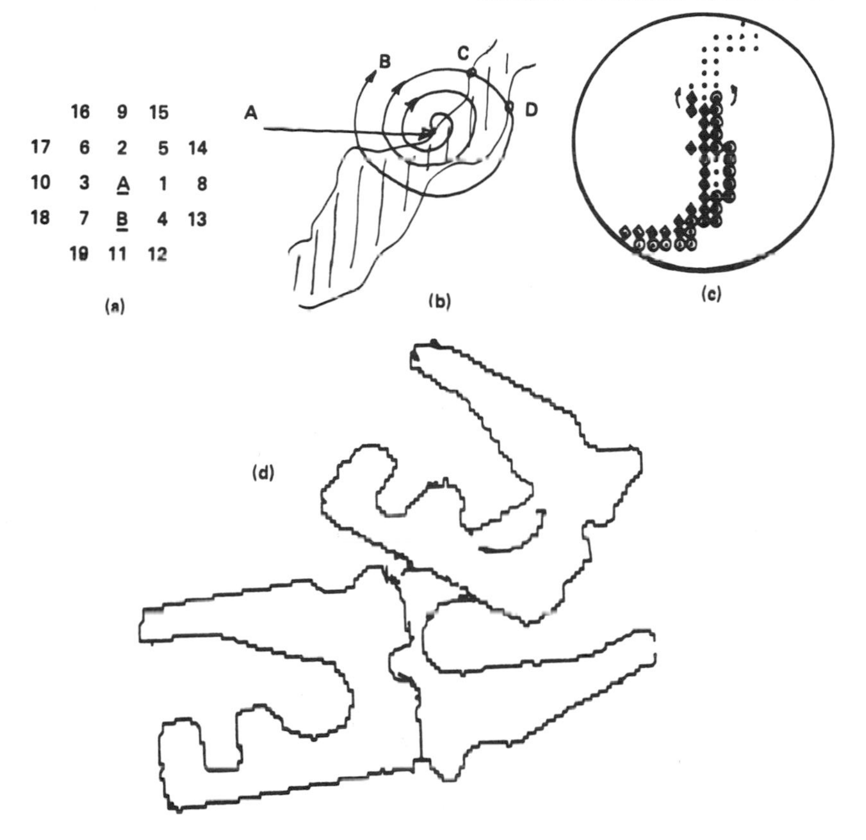

Figure 5. Sequential line-thinning. (a) 19 neighbours of a tracker are ranked by priority. The tracker is A, coming from B. (b) Search of 2 starting positions for complementary trackers on each side of a contour. The picture is scanned (A) until activity is found. Then entry and exit points are searched on a spiral path (B). The complementary trackers are located in C and D. (c) Trackers at work. (d) Contours of Figure 3 after processing

closed loops are found when the trackers return to their starting positions.

Computation versus Use of Look-up Tables

Look-up tables allow fast implementations but computation is necessary where high accuracy is required.

As an example in robotics, consider the problem of curve smoothing. Many strategies have been described for curve smoothing, including polygonal and spline approximation [15, 16, 17], and straight-line and circle-arc fitting [18]. Each of these strategies has its advantages. At the extremes, however, a curvilinear filtering [9] provides the best accuracy (Fig. 6b), and a look-up table conversion (e.g., [4]) leads to the fastest implementation (Fig. 6c).

The former technique takes into account noise and signal statistics in the Fourier domain, and thus is closely related to the least square filter (Wiener). Its peculiarity lies in the fact that curves need not be sampled and part of the processing can be recursively designed.

A curve C is conveniently described by parametrized functions x(s) and y(s):

$$c: \quad \{x(s), y(s)\} \tag{2-2}$$

If x(s) and y(s) are quantized, they become piecewise constant functions. The curve reconstructed from these piecewise constant functions consists of isolated picture elements (pixels). Moving along the quantized curve means repeatedly jumping extremely rapidly from one curve element to the next, and then resting there for a finite amount of "time" (Δs).

In images, spatial variables are more meaningful than time variables. Therefore, the parameter s is usually defined as the curve arc length:

$$ds = \sqrt{dx^2+dy^2} \tag{2-3}$$

Equation (2-3), however, induces a nonlinearity which relates curve amplitude to curve frequency.

Thus the curve spectrum has an upper bound which steadily decreases along the (spatial) frequency axis. Quantization noise, on the contrary, has a constant upper bound. Therefore lowpass filtering ("curve smoothing") is appropriate. The higher the curve signal-to-noise ratio (SNR) required, the lower must the cut-off frequency be set. The filtering is expressed by:

$$x'(s) = x(s) * h(s) \tag{2-4}$$

$$y'(s) = y(s) * h(s) \tag{2-5}$$

where * denotes convolution and h(s) is the impulse response of a lowpass filter (cut-off spatial frequency f_c [cycles per length unit]).

Usually the smooth curve resulting from this filtering has lost some of its own high frequency content as well as some quantization noise.

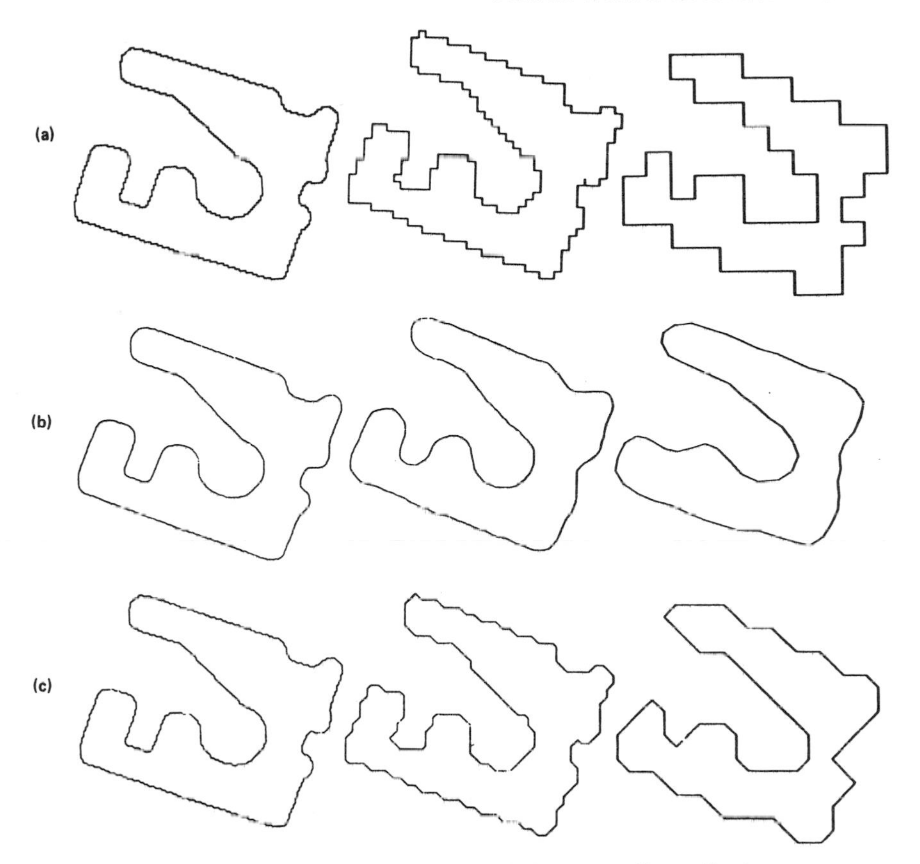

Figure 6. (a) Original data. (b) Smooth curves result from a curvilinear filtering. Filter parameters are chosen differently in each case, depending on noise power. (c) Data are smoothed using the look-up table of Figure 7

High-frequency definition is traded for insensitivity to noise.

In practice, the parameter s is unknown and must be estimated from the noisy curve. However, the smoothed curve allows a better estimation of arc length. The filtering can start again along the improved estimate. Thus the solution may be iterative [9].

The latter strategy, using a look-up table, derives from considerations of local curve distortions. Some particular configurations of contour elements can be replaced by smoother contributions. A look-up table contains the corresponding incremental arc lengths. This technique often turns out to be equivalent to a particular case of the former approach. For example, the table shown in Fig. 7 contains arc increments that could result from computing actual arc increments after curvilinear filtering with a 2-element wide window.

GEOMETRICAL CONFIGURATIONS	LENGTH CONTRIBUTION	ELEMENTARY WEIGHTS
		1
		$\sqrt{2}$
		1.1180
		$1/\sqrt{2}$

Possible geometrical configurations of contour elements and their correspondig elementary weights

Figure 7. Look-up table of arc increments corresponding to smoother configurations

Computer Addressing versus Raster Scanning

Hardware (or a microprogram) allows fast data access, but often lacks flexibility. Simple addressing schemes are required.

A convenient example is offered by polar mapping, a well-known practice proven useful for scene analysis. This technique induces a correspondance between a rectangular coordinate system and an r-α plane. This transformation can be invariant to scene translation when a particular point of the scene defines the origin of the R-α system. Thus, part recognition and orientation estimation are made more easily.

The inherent flexibility of a GPC allows one to map contour elements from the cartesian reference system to the polar plane in random order (Fig. 8). For each element P_{ij} (x, y), the polar coordinates are defined by:

$$\alpha_{ij} = \measuredangle \overrightarrow{OP_{ij}} = \arctan\ (y_j - y_o)/(x_i - x_o) \quad (2\text{-}6)$$

$$r_{ij} = |\overrightarrow{OP_{ij}}| = \sqrt{(x_i - x_o)^2 + (y_j - y_o)^2} \quad (2\text{-}7)$$

where 0 denotes the polar system origin, usually placed on a part centroid.

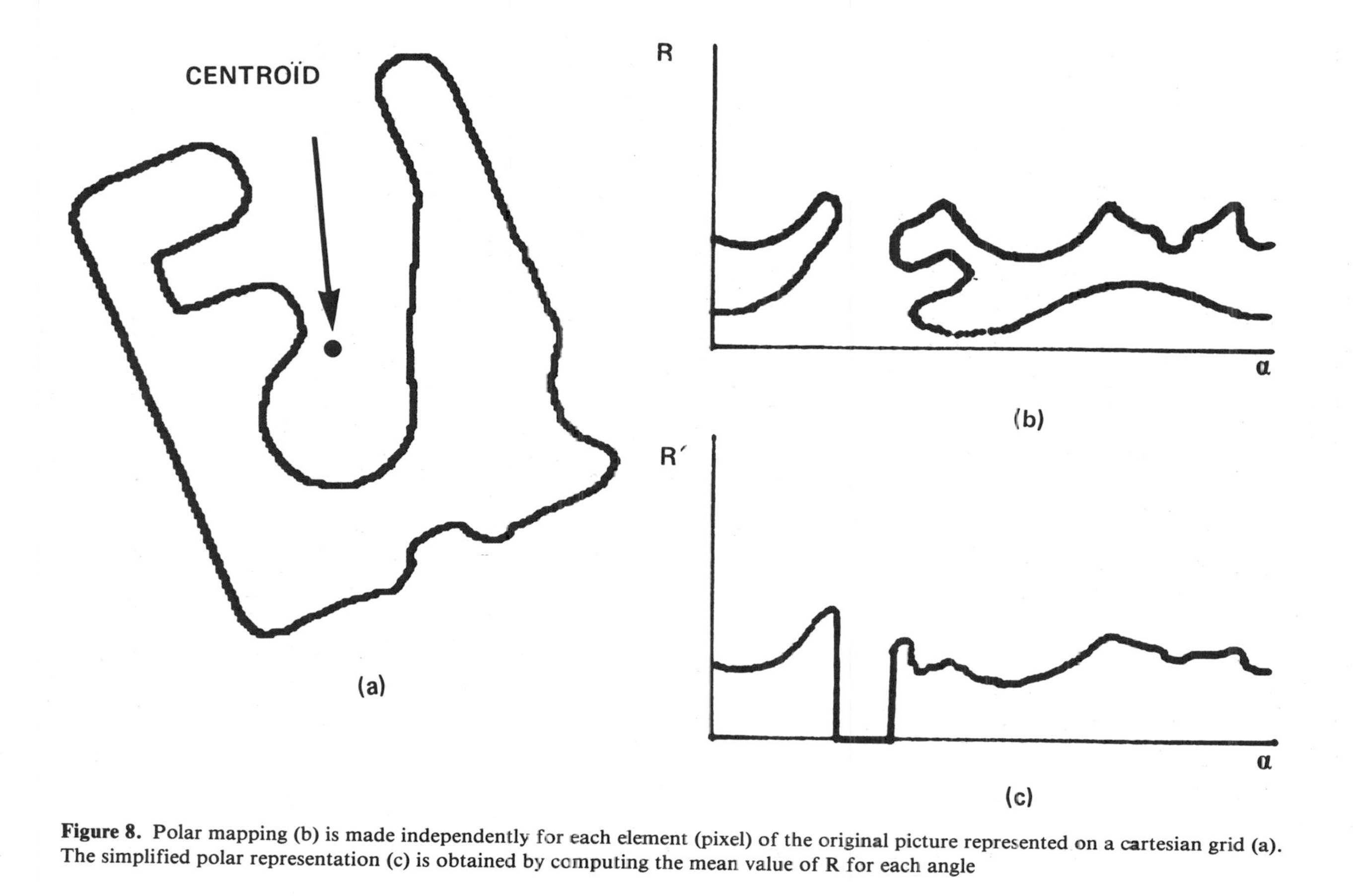

Figure 8. Polar mapping (b) is made independently for each element (pixel) of the original picture represented on a cartesian grid (a). The simplified polar representation (c) is obtained by computing the mean value of R for each angle

The R and α estimation implies arctangent and square-root computations, which are standard features on most GPC's. When specialized hardware is to be designed this technique leads to complex circuit structures. Systematic picture scanning is preferable in this case (Fig. 9). Through the use of a particular scanning, the process is restricted to integer angle values and the corresponding weights associated with the x and y coordinates can be easily tabulated [12].

Experiments show that on a GPC (PDP11/40 computer, Fortran-11 language, and Macro-11 assembler) the former strategy is faster, despite the rather lengthy arctangent and square-root operations. This is due to the flexibility of the GPC, which allows selective data access. On the other hand, the latter method can easily be implemented with SSI and MSI circuits, and has been used for industrial-part recognition and localization (position, orientation, and visible-face identification), performed at the speed of 12 parts per second [12].

Structural versus Correlative Comparisons

Correlative comparisons are easily implemented by specialized circuitry, and lead to fast processing. Structural strategies can reduce the bulk of computation when a GPC is used.

More important, structural techniques can cope better with distortions of all kinds, while correlative comparisons are less sensitive to additive noise [19]. This may sometimes lead to a more decisive criterion than implementation medium considerations.

Industrial-part orienting provides an interesting example of a situation where structural and correlative methods may compete.

Consider again the polar mapping already described. The polar representation $R(\alpha)$ can be made insensitive to part position in the scene. Changing orientation, however, induces an α-delay in the polar plane. If the part is a priori known by its representation $R_r(\alpha)$, a correlative technique can be used. The delay corresponds to the value Θ_0 for which the correlation function $C(\Theta)$ reaches its maximum value:

$$c(\theta) = \int_{2\pi} R_r(\alpha) \times R(\alpha+\theta) \, \partial\alpha \qquad (2\text{-}8)$$

In order to avoid multiplication, the absolute difference function can be used instead of correlation:

$$D(\theta) = \int_{2\pi} |R_r(\alpha) - R(\alpha+\theta) \, \partial\alpha \qquad (2\text{-}9)$$

Taking advantage of the part symmetries (odd and even contributions) a

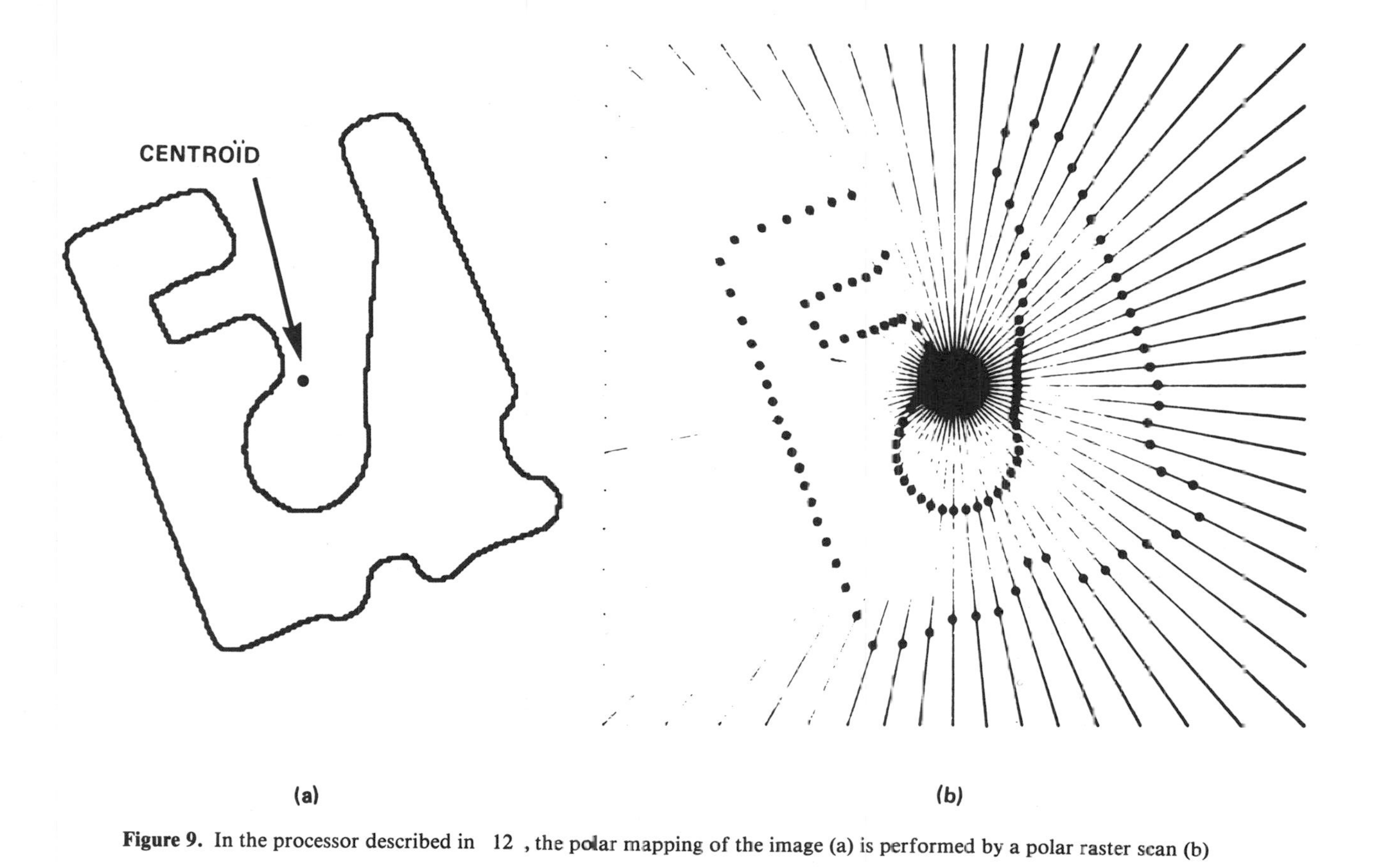

Figure 9. In the processor described in 12 , the polar mapping of the image (a) is performed by a polar raster scan (b)

symmetry axis, and thereby an absolute reference angle, can be more easily estimated [4] (Fig. 10). This angle corresponds to the value Θ_0 for which the odd contributions $S(\Theta)$ of the function $R(\alpha)$ is minimum:

$$S(\theta) = \int_{\pi} |R(\alpha+\theta) - R(-\alpha-\theta) \, \partial\alpha \tag{2-10}$$

This technique requires no multiplication and is particularly valuable when object recognition is combined with orientation estimation. Using correlation, this task would imply N^2 operations per reference. The latter method, besides an initial amount of $N^2/4$ operations for symmetry axis detection, requires only 2N operations per reference.

Template matching and correlative comparison are ideal for hardware because they require a straightforward structure and simple operations. Due to the speed that can be achieved, the relatively high number of computations to be performed is of secondary importance.

On a GPC however, a few "sophisticated" operations are preferable, because simplicity is not rewarded: even the simplest register incrementation needs a relatively long execution time. On a GPC, we have found it useful to utilize structural methods in order to speed up processing. Along a contour, two particular shape details can define orientation [5, 6] (Fig. 11). For instance, an arbitrary vector can originate from a "notch" and reach a "tooth." A "notch" has a concave curvature, while a "tooth" has convex shape. In more general terms, any segment of the contour curvature function is suitable to characterize a vector end.

Under the restrictive assumption that parts are fully visible in the scene, the orientation vector can be optimized during a learning stage. The vector ends, optimal in terms of noise sensitivity and shape ambiguity, are stored and characterized by a curvature function in the appropriate neighborhood. During scene analysis, contour curvature is estimated and compared to stored data. If by this operation the vector ends are recognized, orientation estimation follows.

If parts are partly hidden (e.g., if they overlap), then the orientation vector must be generated from case to case. It can be mapped onto reference data through curvature comparison.

Conclusion

Hints are given for the choice of algorithms specific for implementation either on a general-purpose (micro) computer or on specialized hardware. More valuable are perhaps the examples of image analysis

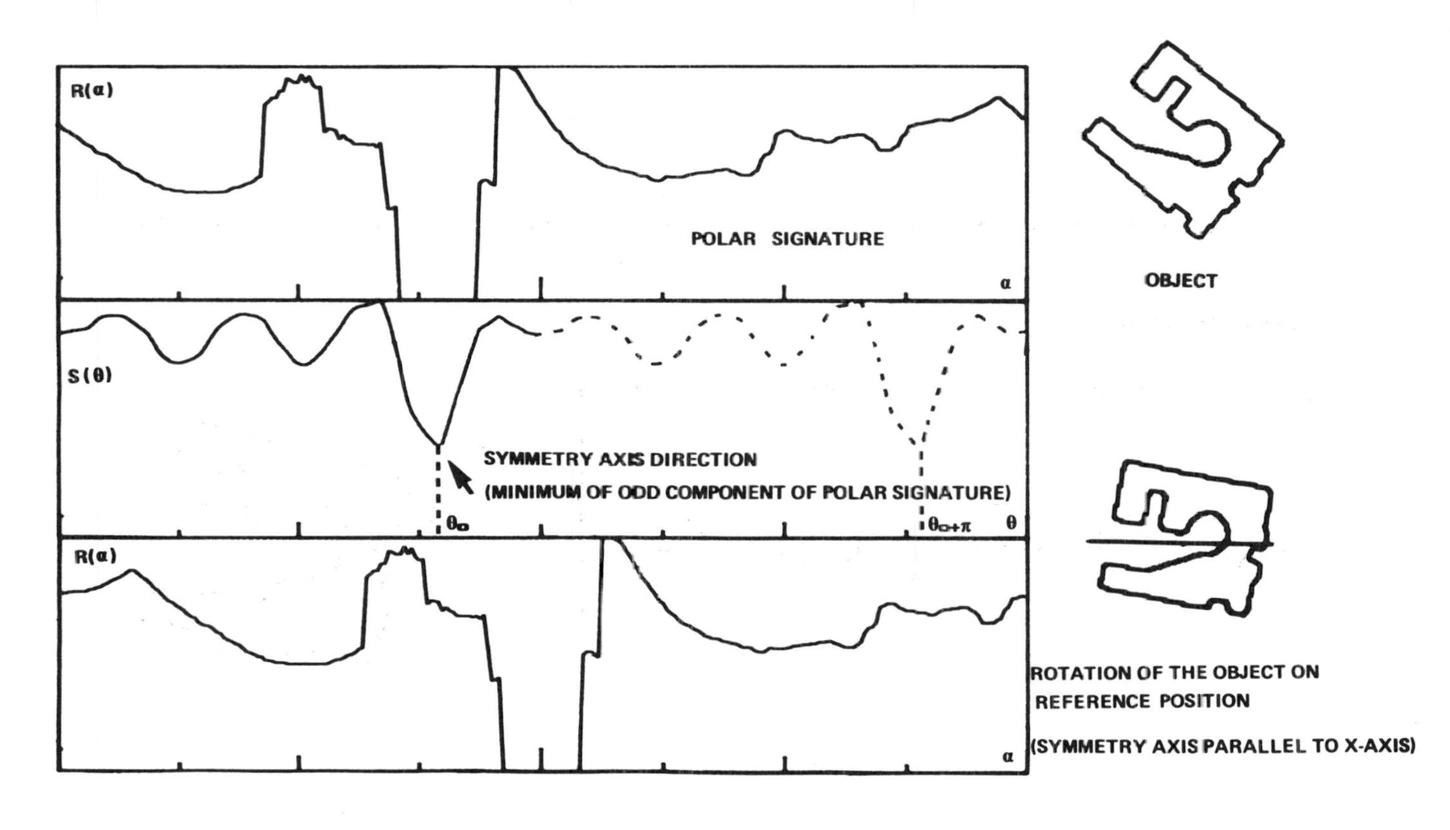

Figure 10. The orientation of the object is defined modulo 180 degrees by the direction of a symmetry axis

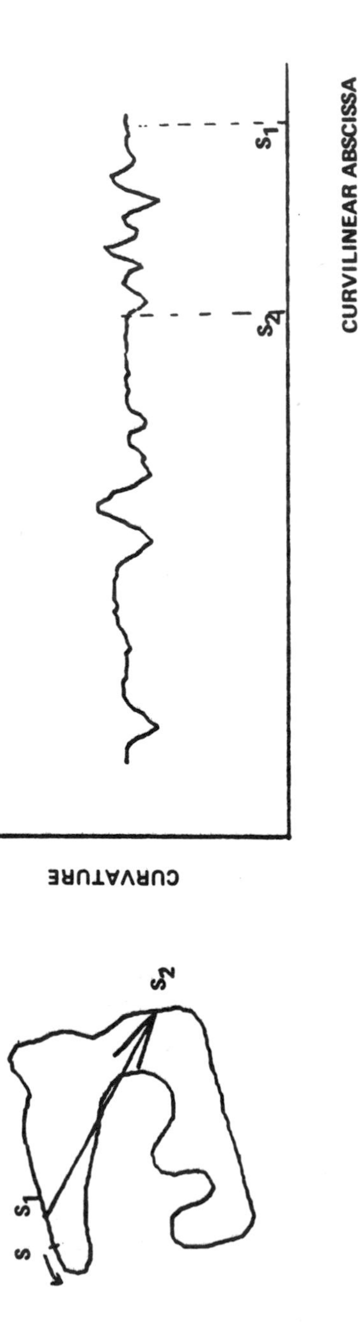

Figure 11. The orientation is defined by an arbitrary vector. A short curvature function is associated with vector ends

strategies that result from our experience in robotics. It should be noted that the trend is toward hardware integration of the most widespread procedures.

Thus, designers can see specialized hardware solve more and more complex tasks, while preserving a very simple structure at board level. "Digital design" means more and more "system design."

The GPC programmer, on the other hand, is often torn between two mutually exclusive aims:

To develop and tune up algorithms that are optimal on the GPC.

To develop and refine on a GPC algorithms that will prove optimal when implemented with specialized circuitry.

Acknowledgment

The authors wish to thank Prof. F. de Coulon for many helpful discussions and stimulating remarks.

References

1. Business Week, "Robots join the labor force," June 1980, pp 62–76.
2. IEEE Computer, special issue on machine perception, May 1980, pp 11–63.
3. F. de Coulon, P. Kammenos, "Polar coding of planar objects in industrial robot vision," *Neue Technik,* No. 10, 1977, pp 663–671.
4. P. Kammenos, "Performances of Polar Coding for Visual Localisation of Planar Objects," Proc. 8th International Symposium on Industrial Robots, Stuttgart, W. Germany, pp 143–154, May–June 1978.
5. J. D. Dessimoz, "Visual Identification and Location in a Multi-object Environment by Contour Tracking and Curvature Description," Proc. 8th Intern. Symp. on Industr. Robots, Stuttgart, Germany, pp 764–777, May–June 1978.
6. J. D. Dessimoz, M. Kunt, G. H. Granlund, J. M. Zurcher, "Recognition and handling of overlapping industrial parts," 9th Intern. Symp. on Industr. Robots, Washington, USA, March 13–15, 1979, pp 357–366.
7. P. Kammenos, "Extraction de contours en traitement électronique des images I: Principaux operateurs de traitement," *Bull. de l'ASE,* Zurich, No. 11, Juin 1979, pp 525–531.
8. J. M. Zurcher, "Extraction de contours en traitement electronique des images II: Processeur specialise pour signal video," *Bull. de l'ASE,* Zurich, No. 11, Juin 1979, pp 532–536.
9. J. D. Dessimoz, "Curve smoothing for improved feature extraction from digitized pictures," *Signal Processing,* 1, No. 3, July 1979, pp 205–210.
10. J. D. Dessimoz, "Specialised edge-trackers for contour extraction and line-thinning," *Signal Processing,* 2, No. 1, Jan. 1980, pp 71–73.

11. J. D. Dessimoz, "Sampling and smoothing curves in digitized pictures," Proc. 1st EUropean SIgnal Processing COnference, EUSIPCO-80, Lausanne, Sept. 16–19, 1980, pp 157–165.
12. J. M. Zurcher, "Conception d'un système de perception visuel pour robot industriel," Comptes-rendus des Journees de Microtechnique, Ecole Polytechnique Federale de Lausanne, Sept. 1978.
13. H. Blum, "A transformation for extracting new descriptors of shape," Models for the perception of speech and visual form, W. Wathen-Dunn, ed., MIT Press, 1967, pp 362–380.
14. C. Arcelli, L. Cordelli, S. Levialdi, "Parallel thinning of binary pictures," *Electronic Letters* (GB), 11(7), 1975, pp 148–149.
15. T. Pavlidis, "Polygonal Approximations by Newton's Method," *IEEE Transactions on Computers,* vol. C-26, no. 8, Aug. 1977.
16. J. E. Midgley, "Isotropic Four-Point Interpolation," *Computer Graphics and Image Processing,* vol. 9, pp 192–196, 1979.
17. H. G. Barrow and R. J. Popplestone, "Relational Descriptions in Picture Processing," *Artificial Intelligence,* vol. 2, pp 377–396, 1971.
18. F. L. Bookstein, "Fitting Conic Sections to Scattered Data," *Computer Graphics and Image Processing,* 9, pp 56–71, 1979.
19. M. Berthod, J. P. Maroy, "Learning and syntactic recognition of symbols drawn on a graphic tablet," *Computer Graphics and Image Processing,* vol. 9, pp 166–182, 1979.

Chapter 3

Comparison of Five Methods for the Recognition of Industrial Parts

J. POT

Jet Propulsion Laboratory, Pasadena, USA. Currently at
Laboratoire d'Automatique de Montpellier, France

P. COIFFET

Laboratoire d'Automatique de Montpellier, France
U.S.T.L.

P. RIVES

Bell Northern Ltée, Verdun, Canada. Currently at
Laboratoire d'Automatique de Montpellier, France

Abstract This paper describes algorithms for recognition of industrial parts. We assume that the camera is fixed. This vision system will be used for a robot. Five different methods are compared with regard to their efficiency, speed, and flexibility of utilization. As far as the sequential Bayesian classifier is concerned, the utilization of a trapezoidal law seems a little better than the other methods. A processor using these five methods has been completed.

Introduction

Some computer vision systems already exist (1), and their utilization will grow. For robots that have to manipulate objects, vision is a very powerful method of carrying information about these objects.

This information is extremely complex. An image is composed of a great number of pixels. Much research has been done to find methods to interpret images. Nevertheless, the vision problem has not been mastered. Some systems are working under severe constraints.

In this article, we describe a system working under the following constraints: objects are stationary and non-overlapping; the camera is situated above the working plane and its optic axis is perpendicular to the plane; the lights are calibrated and there is reasonable contrast between objects and background so that we can get a binary image using a fixed threshold.

Given these conditions, which are usually found in the vision systems working in industry, we are trying to reduce these conditions (2), this article describes some methods of object recognition and the learning accomplished by the system. The goal is the achievement of the system which:

1. Uses a very simple method of learning (learning by showing objects to the camera, for example);
2. Demonstrates a good success rate (near 100%); and
3. Utilizes fast algorithms (recognition is followed by a robot action).

We compare five original variations on two categories of classical methods.

Choice of Features and Errors

We assume that, using 2-D techniques, there are as many object models as there are different equilibrium faces.

When we put an object on the working plane, the variations of its image have the following possible causes:

1. A change due to the variation of lighting of the scene.
2. The variations due to the discretization of the image.
3. The variations in the position and the orientation of the objects on the working plane. These errors are due to parallax.

We can either take the hypothesis or show experimentally that with our constraints the major cause of errors is due to discretization. In any case, we can take account of all the possible causes for errors in the same model (3).

The first step will be to separate the objects in the binary image using a segmentation algorithm.

Features of various types can be used. These can be topological, geometric, or analytical (4). The choice of the best features depends on the initial set of objects. Nevertheless, it is interesting to keep a general aspect to recognition.

We have studied the behavior of some features on geometrical objects, that is, 10 different equilibrium faces. These features are:

1. The area of the objects (SURF).
2. The maximum, minimum, and average distance from the center of inertia of the image to the edge (RMAX, RMIN, RMOY).
3. The perimeter of the image (i.e., the number of points of the edge).
4. The corrected perimeters. Each point on the edge receives a weight depending on its situation. In the first method, we use a weight of $\sqrt{2}$ for each diagonal point (PER2). The second method uses a more sophisticated method of weighting (PER3) (5).
5. The first four combinations of the image moments used by Dudani (6) with some variations. These are:
   - 5-1: Moments computed on the full image (MP1, M1, M2, M3).
   - 5-2: Moments computed on the edge, using the center of inertia of the full image.
   - 5-3: Moments computed on the edge, using the center of inertia of the edge.
   - 5-4: Moments computed on the edge, using the method of weighting used for PER2.
   - 5-5: Moments computed on the edge, using the method of weighting used for PER3.
6. The first seven coefficients of the Fourier transform of the distance from the center of inertia to the edge.
7. The dimensions of the minimal rectangle bounding the image.

An interesting parameter must have the following properties:

1. It must be stable with respect to the variations of the images of the same object.
2. It must be distinguishing.
3. It must be fast to compute.

Examples of the variations of some parameters are shown in Figure 1. The values of the parameters are functions of the angle ψ. ψ is the angle of rotation of the object around the optical axis of the camera.

We have made an analysis of correlations and also a discriminant factorial analysis using these features. We also can evaluate the time of computation of these features. These enable us to keep only six features. These are SURF, DMAX, MP 1, M1, M2, M3.

This choice is not universal. For example, we do not use the dimensions associated with the holes of the objects. Nevertheless, this set of features has been efficient enough for the experiments we made after that.

First Method, Using Linear Discriminant Functions

Many algorithms exist that give the coefficients of linear functions which separate two classes of objects (7, 8). We don't use the gradient method (9) because it doesn't converge very fast. We use the method of discriminant factorial analysis.

The principle is to find a function which divides a given set of classes into two subsets, using n features. If the separation is not efficient enough, we introduce a new parameter. The steps of the algorithm are shown below:

Initialization

We choose the first feature. The initial set consists of all the initial classes.

Iterations

Step 1: For one set of classes CK1,. . . ,CKn, we select 2 classes CKO and CKl, using a criterion.

Step 2: We compute the linear discriminant function that separate these two classes, using the selected features.

Step 3: Using a criterion, we verify that these two classes are separated. If not, we go to step 6.

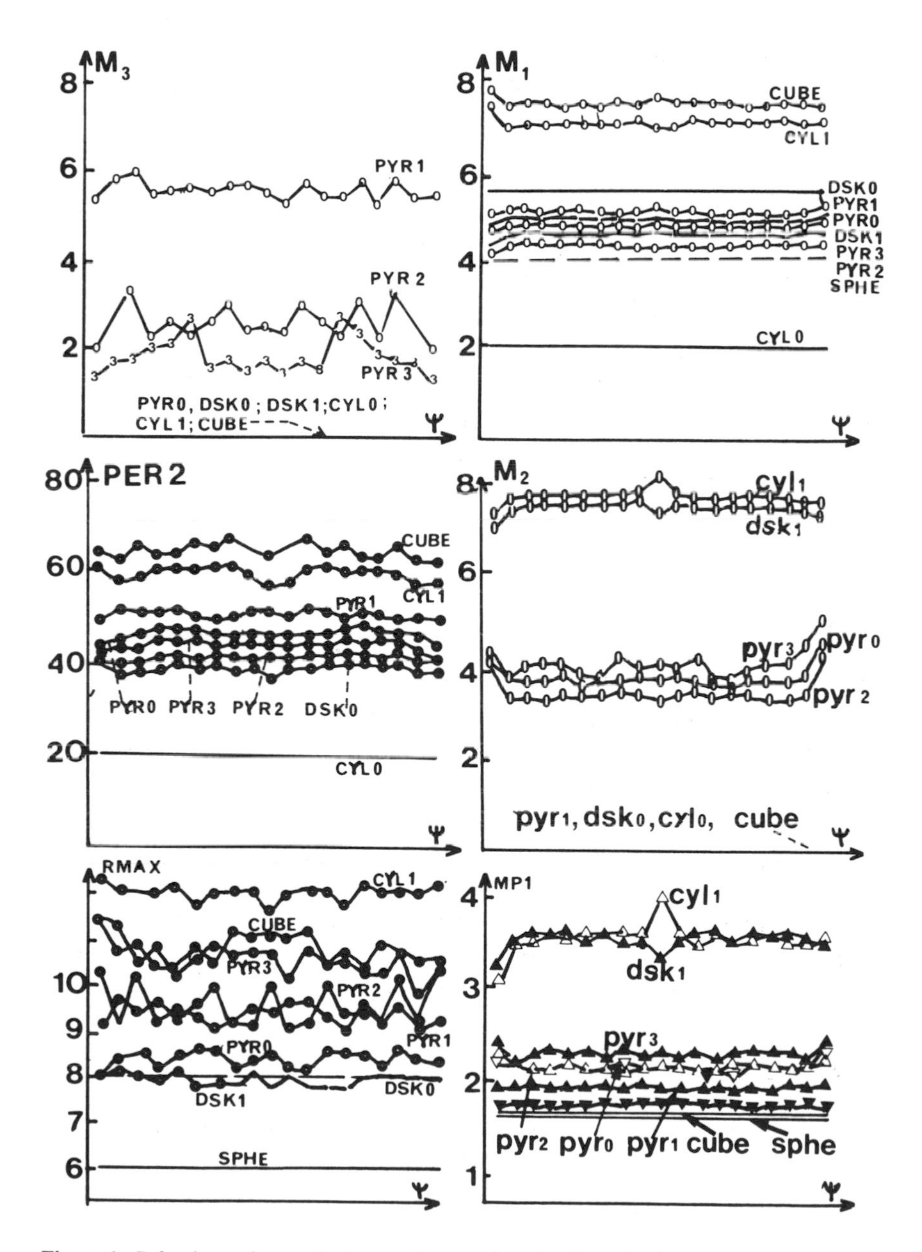

Figure 1. Behaviour of some features with regard to the discretization of a sphere, a cube, a disk (2 equilibrium faces), a cylinder (2 e.f.), and a pyramide (4 e.f.)

Step 4: Each class of the initial set CKl,. . . ,CKn is attributed to one of the two subsets, according to its position with resect to the separating hyperplane. If a class straddles the hyperplane, it is attributed to both subsets.

Step 5: If all the subsets have only one class, the process is completed. This means that each class can be described by a set of linear discriminant functions. If not, we go to step 6.

Step 6: If it is possible, we now select a new subset having more than one class and go to step 1. If it is not possible, that means that we cannot separate the remaining sets with the introduced features. So we introduce a new feature, select a set, and go to step 1.

Choice of the Criterion

For step 1

We compute a hierarchy in the selected set of classes. The last two classes introduced in this hierarchy are the selected classes CKO and CKl. The hierarchy is computed using the Mahalanobis distance between the center of inertia of the classes.

For step 2

The coefficients of the linear discriminant function are computed so that they maximize the ratio of the interclass variance to the intraclass variance, regardless of the selected features (10).

For step 3

Two classes CKO and CKl are separated if all the samples of one class are on the same side of the discriminant hyperplane.

For step 4

A class straddles the discriminant hyperplane if there are samples of this class on both sides of the hyperplane, or if the projection of the center of inertia of the class on the discriminant axis is between the projections of the centers of inertia of the two classes CKO and CKl.

When this algorithm is completed, we have a discriminant tree (Figure 2a, b). This tree is represented by cells. Some cells characterize the nodes of the tree. Each of these cells contains five values (i1, i2, i3, i4, i5). For the node K, i5 is the number of coefficients of the linear discriminant function. Other cells contain the coefficients CKi and the threshold SK of this function. Thus, for a given image, if we compute the values of the features P1, . . . ,Pn, we can compute the value of the question at the node K, that is:

$$Q_K = \sum_{i=1}^{i_5} (CKi) \times (Pi) - SK \quad (3\text{-}1)$$

i1 (i2) is the number of the recognized object if QK is positive (negative), and if the branch of the tree indicates an object. If not, i1 (i2) is equal to zero, and i3 (i4) is the number of the following node in the tree.

Discussion of This Method:

This method allows us to reduce the theoretical number of required nodes. In the case of 23 objects, the algorithm uses 32 nodes. The theoretical method requires 253 nodes. This method is also sequential, that is, we are not obliged to compute all the features, but only those necessary for a specific step. Thus, once the tree is computed, the recognition is very fast. If the external conditions do not vary too much, the results are very good. Furthermore, this algorithm can easily be implemented in hardware. Nevertheless, there are some drawbacks to this method. If we add or subtract an object of the initial set, we have to compute a new tree. The coefficients of the tree are quite difficult to generate in real time. These can be obstacles for utilization of the method in factories.

Second Method: Utilization of Distances

Using the learning samples, we can compute for each class K the average gkj and the standard deviation kj of the J selected features.

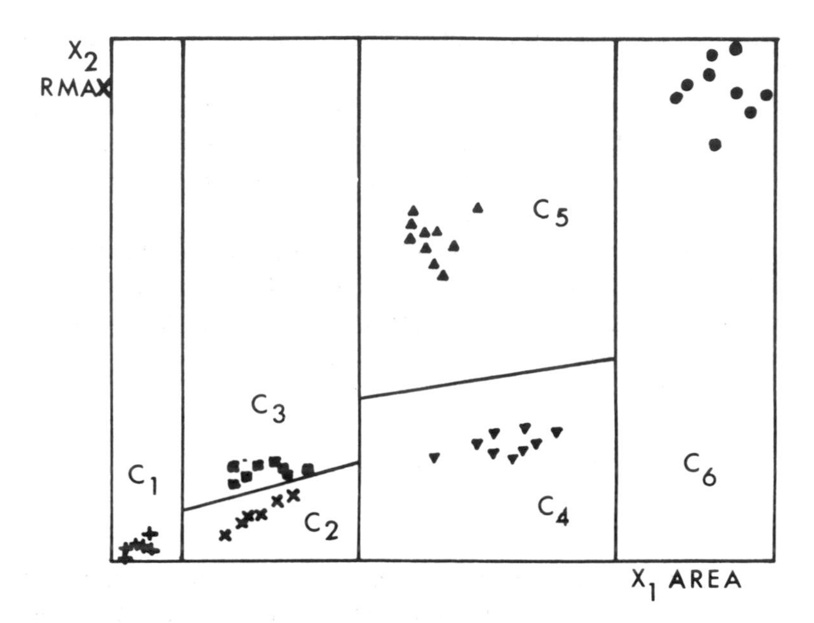

Figure 2a. The spatial repartition of the learning samples of 6 classes

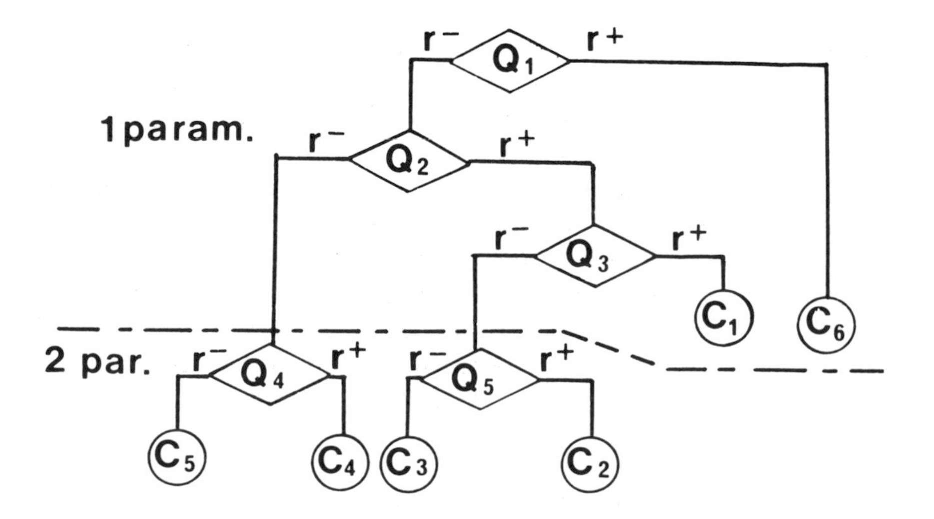

Figure 2b. The tree generated by the learning algorithm

When we get a new image, we can classify it as follows:
1. We first compute the jth feature xj.
2. We compute the distance $dkj = |xj - gkj| / \sigma kj$ for all remaining classes. If dkj is greater than a threshold, the class k is eliminated.
3. We have then three alternatives:
   3.1 dkj > threshold for all k. The object is then unknown.
   3.2 dkj < threshold for only one class k0. The object is recognized.
   3.3 dkj < threshold for more than one class. We introduce a new feature, and go to step 1.

If all the features have been introduced, and more than one class still remains, we can compute a distance of the object to all the remaining classes.
For example:

$$SK = \sum_{j=1}^{J} dkj = \sum_{j=1}^{J} |xj - gkj| / \sigma kj \qquad (3\text{-}2)$$

The choice of the threshold is very important. It is based on a Gaussian model and represents a percentage of the distribution.

This classifier is very simple and very fast. It not only attributes an object to a class, but also indicates the ambiguous cases, so that other sensors can be used to solve this ambiguity (a moving camera, for example).

Methods Using the Sequential Bayesian Classifier (11)

We assume that the features are statistically independent. The problem is to estimate the probability densities regardless their simplicity and their efficiency. The three estimators are a trapezoidal law, the Gaussian law, and histograms estimated by integer count.

Figure 3 gives some examples of these estimations for the surface of 10 different objects. (The objects are parts of an engine.)

The trapezoidal law is represented by four coefficients x1, x2, x3, and x4. These coefficients are chosen so that:

$$x1 = gkj - \alpha\,\sigma kj \qquad x2 = gkj - \beta\,\sigma kj \qquad (3\text{-}3)$$
$$x3 = gkj + \beta\,\sigma kj \qquad x4 = gkj + \alpha\,\sigma kj \qquad (3\text{-}4)$$

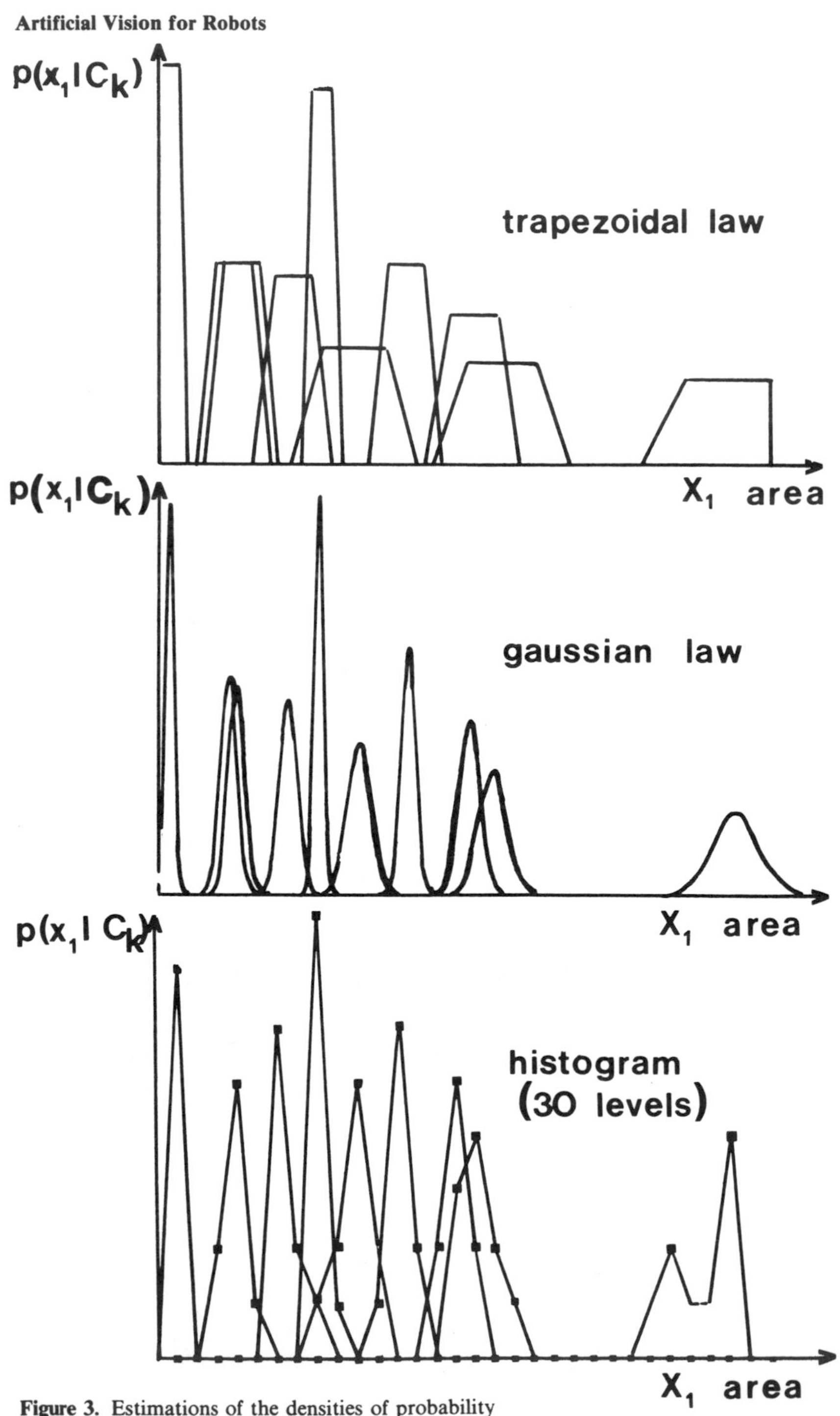

Figure 3. Estimations of the densities of probability

and α and β are such that 90% of the learning samples are between x2 and x3, and 99% between x1 and x4, if we use a Gaussian model.

Normal law is given by the formula:

$$(\wedge / \sqrt{2\pi}\sigma kj)\exp(-(xj - gkj)^2/\sigma kj^2) \quad (3\text{-}5)$$

where gkj and σkj are the average and standard deviation of the jth feature for the class k.

The histograms are computed using a discretization of the field of variation of the features in L parts. The choice of L is very important. Some tests were made using L between 10 and 30.

If 1Nk is the number of learning samples of the class k, and NKj1 the number of instances in which the value of the jth feature belongs to the 1th interval. The probability p ((x1/Ck) is given by Nkjl/Nk.

Thus the recognition algorithm is as follows:

Initialization:

We first compute the value of the first feature of the object.

For each class k, we compute the a posteriori probabilities p(k/x1), using the formula:

$$p(k/x1) = (p(k)p(x1/k))/\sum_{K} p(k)p(x1/k) \quad (3\text{-}6)$$

If p(k/x1) is greater than a threshold (the threshold is 0.9), the object is recognized. If not, we introduce a new feature. The jth step will be:

1. Computation of the values xj of the jth feature.
2. Computation of the probabilities for the remaining classes, using the formula:

$$p(k/x1,..,xj) = p(k/x1,...,xj-1)p(xj/k)/ \sum_{K} p(k/x1,...,xj-1)p(xj/k) \quad (3\text{-}7)$$

3. If p(k/x1,..,xj)is greater than the threshold for a class k0, the object is recognized. If not, we introduce a new feature and go to step 1.
4. If all the features have been selected, and there is still an ambiguity, we have two alternatives:

4.a) We can choose the class for which p(k/x1,..,xj) is a maximum.

4.b) We can resolve the ambiguity using other sensors.

Some Results with the Bayesian Classifier

There are 23 test objects. They are parts of an old motorcycle engine. The learning file is constituted with about 15 samples for each object. The threshold is 0.9, and we choose the following order of computation for the features: SURF, RMAX, MP 1, M2, M3. We can choose an optimal order for these features, using the information theory. We did this, but finally, we preferred to choose a suboptimal but fixed order.

We chose three kinds of errors to assess the quality of the density of probabilities:

a) The big errors. These occur when the algorithm recognizes an object with a good threshold, but there is a mistake.

b) The little errors. This is the case when there is still an ambiguity (i.e., there is no class for which p(k/x1,. .,xj) is greater than the threshold), and the class k for which p(k/xl,. .,xj) is maximum is not the right class.

c) The "not very well classified objects." This occurs when there is still an ambiguity, but the class k for which p(k/xl,. .,xj) is a maximum is the right class.

Error a) gives a measure of the reliability of the algorithm, errors b) and c) give a measure of its accuracy.

The following results were obtained using about 200 samples:

Using the normal law: a) 4%, b) 0%, c) 1%

Total: 5%

Using the trapezoidal law: a) 0%, b) 1%, c) 2%

Total: 3%

Histograms of probability: if L = 10, 40% of the cases are ambiguous. If L = 30, there are 5.5% ambiguous cases.

Nevertheless, we cannot take L too large, for two reasons. First, the memory size used to store these probabilities is limited.

This size is given by S = J L K. If L = 30, S = 4140.

Also, the number of learning samples increases with L, in order to have a good estimation of the histograms.

Also, this method assumes that we know a priori the minimum and maximum of the values of the features. This assumption can be a problem if we add to the initial set a new object whose features are not between the initial values of the extremes.

Thus the trapezoidal law seems to be a little better than the other methods.

Adaptation of the Models: Supervised Recognition

We used a learning file to compute the coefficients used in the algorithms. We can also have supervised adaptation. This latter can be very usful if the external conditions are changing, or if we add a new object to the initial set.

When the recognition system is in this mode, the operator must indicate the real name of the object shown to the camera. The models of the object can then be adjusted. This adjustment can be made only for the last four algorithms. Indeed, the coefficients of the linear discriminant functions cannot be adjusted easily.

Adaptation of the Histogram

The probabilities are adjusted as follows. If the value of the jth feature for the class k is $xj(m_0)$, we have:

$$Pn+1(xj(m)/k) = Pn(xj(m)/k)/(1+\gamma) \qquad \text{if } m \neq m \tag{3-8}$$

$$Pn+1(xj(m_0)/k) = (Pn(xj(m_0)/k)+\gamma)/(1+\gamma) \qquad \text{if } m = m_0 \tag{3-9}$$

The value of γ allows us to choose the speed of the adaptation.

Adaptation of gkj and σkj

The average g and the standard deviation σkj are adapted using the following formulas:

$$g' = (\alpha+x)/(\alpha+1) \tag{3-10}$$

$$\sigma'^2 = ((\alpha+1)\sigma^2+(g-x)^2)\alpha/(\alpha+1)^2 \tag{3-11}$$

Where g' and σ' are the adapted values of g and σ, x is the value of the feature, and α is a coefficient which allows us to choose the speed of the adaptation.

Conclusion

There is a big difference between methods using the linear discriminant functions and the others. The coefficients cannot be adapted easily using this method. Nevertheless, it can be used for specific problems, if the set of objects is fixed.

The use of histograms is the more theoretically justifiable

method. It does not require a priori knowledge of the density pattern. Nevertheless, some problems must first be solved. The first is to find a good value for the number L of discretization intervals, regardless of the memory size and the number of observations needed for good estimation of the histograms. Also, the extremes of the feature variations must be known a priori.

The parametric laws are well behaved, with a little preference for the trapezoidal law. They are very easy to use and very flexible:

The initialization of a new object requires only one single sample.

The adaptation of the models can be done using either an automatic mode or a supervised mode.

These algorithms can make decisions of non-recognition. They can also indicate that there is an ambiguity between different objects. This capability can be very useful, especially in robotics, for generating a decision to get more information with other sensors.

An autonomous vision processor having these properties has been designed. It uses a Reticon camera with a 100 × 100 photodiode array.

We are now studying the collaboration between this fixed camera and a moving one. The moving camera can solve the cases when there is an ambiguity.

References

1. For instance: Optomation II (General Electric), OCTEK 4200, Opto Sense Visual Inspection System (Copperweld), VS110 (Machine Intelligence Corporation).
2. P. Rives. Utilisation d'une caméra solidaire de l'organe terminal d'un manipulateur dans une teache de saisie automatique. Thèse 3° cycle, Montpellier, 1981.
3. J. Pot. Localisation et Reconnaissance de pièces par un système de vision autonome. Rapport interne, L.A.M. Montpellier, juillet 1981.
4. P.Coiffet. *Les Robots - Tome 2: Interaction avec l'environnement.* Editions Hermès, 1981.
5. J.D. Dessimoz. Curve smoothing for improved features extraction from digitized pictures. *Signal Processing* 1, pp. 205–210 (1979).
6. Dudani. Moments method for identification of the 3 dimensional objects from optical sensors, Ohio University thesis, 1971.
7. G. Perennou. Contribution à l'étude des discriminateurs: calcul et optimisation. Thèse, Toulouse, 1968.
8. S. Castan. Contribution à l'étude des problèmes de reconnaissance de formes: optimisation des discriminateurs. Thèse, Toulouse, 1968.
9. Briot. La stéréognosie en robotique - Application au tri des solides. Thèse d'etat, Toulouse, 1977.

10. J.M. Romeder. Méthodes et programmes d'analyse discriminante *DUNOD* (1973).
11. C. Roche. Information utile en reconnaissance des formes et en compression des données. Application à la génération automatique de systèmes de reconnaissance optique et acoustique. Thèse d'état, Paris 7,1972.

Chapter 4

Syntactic Techniques in Scene Analysis

S. GAGLIO
P. MORASSO
V. TAGLIASCO

Istituto di Elettrotecnica, Genova, Italy

Abstract Linguistic concepts provide a powerful means to handle the complexity of scenes. To this goal, we propose a hierarchy of scene descriptors: edges, lines, figures, and objects (we restrict ourselves to two-dimensional scenes). A syntactic-stochastic procedure is described for the recognition and classification of lines. Finally, a computational model (a "pandemonium" of "demons") is briefly described which could be used to implement some current ideas on human perception.

Introduction

The degree of complexity of manipulation processes in human and advanced robots is such that it is not possible to think in terms of direct visuo-motor transformations, such as those present in low-level animals or simple industrial robots.

An alternative to the inextricable difficulty of the direct approach, a linguistic approach first of all provides a reduction of complexity by passing from the world of processes to the world of abstract symbolic descriptions. However, abstraction alone is not sufficient to define and justify a language. A second requirement is to provide a sufficient representative power to preserve much of the "richness" of the visuomotor world.

In a linguistic approach, this power is obtained by adding to the abstraction capability (linked to the definition of "primitive features") a generative capability, that is, a system of rules which allow us to assemble, disassemble, and compare symbolic structures.

A competence theory for the verbal language [1, 2] is a model, in terms of concepts and rules, of the mental structures which underlie speaker-listener communication, independently of the processes which generate specific streams of phonatory movements/sound/auditory signals.

Similarly, a competence theory for the visuo-motor language is a model of the primitive notions and structuring rules which underlie manipulation. In this paper we restrict ourselves to the visual side, with the further simplification of considering only two-dimensional objects. In particular, we propose a set of notions and rules for the description of static, planar scenes and we discuss also some specific syntactic algorithms which can be used for dealing with such a structure (that is, the performance level of analysis which corresponds to the competence level).

One might point out that the analysis of static, planar scenes does not require such a sophisticated linguistic formulation and that simple heuristic algorithms such as polar coding [3, 4] are more straightforward. This is certainly true, and the linguistic approach cannot compete with ad hoc systems in the planar case. However, for three-dimensional scenes and for integrated visuo-motor cognitive manipulation systems the linguistic approach, in our opinion, is the best candidate in the long run. As a consequence, the contribution of this paper is quite preliminary, with the main purpose of helping to define concepts and algorithms in a simple case.

Ambiguity and Assumptions

Uncertainty is intrinsic at each level of analysis. At low levels, however, uncertainty is basically due to "noise" in the data (that is, in the image), whereas at high levels uncertainty takes the form of the intrinsic ambiguity of trying to characterize the world of objects by means of an opto-sensitive detector.

Therefore, the type of analysis appropriate at low levels is data-driven, whereas at high levels it must be conceptually driven, that is, it is necessary to enrich the insufficient data with assumptions which, on the basis of the verification of relations among the available data, can allow us to infer the existence of missing elements or to choose among different possible interpretations of the same visual data.

In a conceptually-driven organization, it is possible to use context-free grammars at the different levels of analysis even if the nature and the ambiguity of the vision task would require the power of a context-sensitive analysis. In fact, in a conceptually-driven organization context-free grammars play only a reference role which has to be coupled to the verification of context-dependent assumptions.

The structure of this paper reflects the duality between data-driven and conceptually driven analysis. First, we discuss edges and lines.

Then, before dealing with objects and figures, we propose a formal description of a scene which is a conceptual framework for high-level scene analysis.

Acquisition of Images and Extraction of Features

In the image of a scene, the information about the objects is "buried" in the spatial distribution of the gray levels (or of the color levels). In order to make explicit what is already implicit in the image (that is, to interpret and understand the scene) it is necessary to extract some features present in the image and to look at their spatial organization.

Among the many different features which can be extracted from the image of a scene, we shall concentrate on edges, that is, we shall look for contrast discontinuities in the image [5].

The choice is motivated, on one side, by the fact that line drawings are adequate representations of many scenes and, on the other, by our opinion that other types of features (for example, textures, illumination gradients) can be used in a pre-processing of the image which enhances significant discontinuities, making edge detection and edge processing algorithms applicable.

In the context of this paper we shall make the following hypotheses concerning the initial phase of analysis:

(i) An image of a scene can be appropriately acquired in digital form, as a spatial distribution of picture elements (pixels) over an optoelectronic sensor,

(ii) The pixels of the digitized image can be distributed among a number of receptive fields (Fig. 1).,

(iii) Each receptive field can be associated with a primitive operator which detects the presence of contrast discontinuities in the field (in particular, we shall consider "edge elements," where an ideal edge can be defined as the straight boundary between two adjacent areas characterized, within the receptive field, by a step change in contrast).

The possibility of applying syntactic techniques for exploring the organization of the picture stems from the fact that, if the possible orientations of the edge elements are discrete and small, it is then possible to label them with the symbols of an alphabet.

In our case, we have discretized the orientation into 16 possible levels and, accordingly, we have used an alphabet (a,b,c, ...,p) where a is associated with the vertical orientation and the other symbols identify progressive rotations (pi/16) in the clockwise direction.

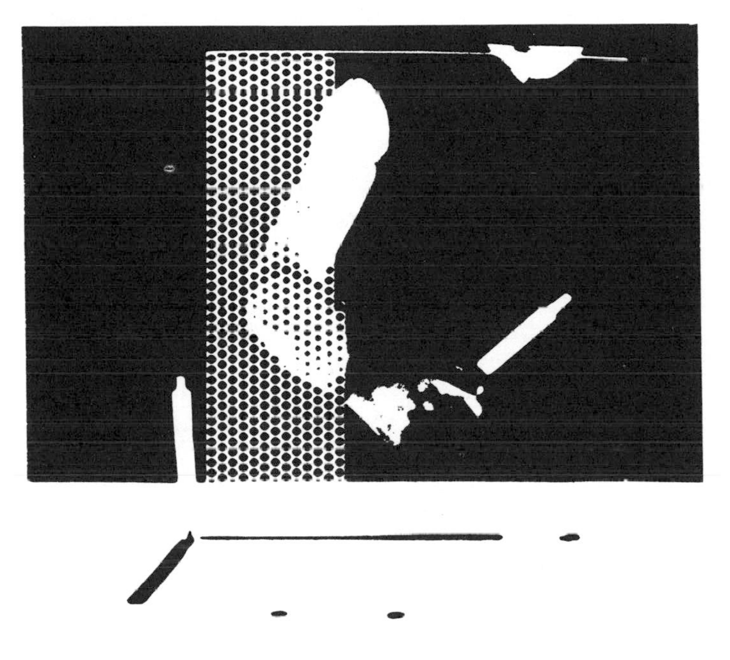

Figure 1. Pixels of digitized image can be distributed among receptive fields

Edges

Two edge detection algorithms were developed in our laboratory for performing some experimentation with scene analysis.

The two algorithms were inplemented for the PDP 11/34 minicomputer and were used to process digitized images on a TV camera by means of a Tesak VD501 Image System (interfaced with the PDP 11/34 by means of a digital parallel I/O interface DR11C), which stores an image with a resolution of 512 × 512 pixels, each of them quantized with 8 bits.

Hueckel Operator

This operator has been proposed by Hueckel [6] for a continuous distribution of intensity F (x,y) over a unit circle, which is optimally fitted by an "edge," that is, a bidimensional step function.

In our computer implementation of the Hueckel operator, we expressed the operator by means of explicit formulas, which weighted the

8 × 8 pixels of a receptive field [7].

The resulting processing time of the image (without a floating point unit) is about 7 minutes.

Gradient Operator

Another operator was developed in our laboratory [7] which uses a 5 × 5 receptive field and is based on the computation of the intensity gradient (by means of a very simple 5 × 5 mask). If an edge is detected, a smaller 3 × 3 mask is used to refine the position of the edge.

For the class of images studied, the computation time is about 45 seconds.

Summing up, the initial phase of feature extraction transforms the image of Fig. 1 into the image of Fig. 2. From the computational point of view, the transformed image is expressed as a set of LISP-like descriptors, according to the following structure:

```
(<edge name> (CLASS EDGE)
             (POSITION <x-coord> <y-coord>)
             (ORIENTATION <angular label>))
```

Lines

A primitive line can be defined as a collection of adjacent edges limited between two vertices, where a "vertex" is either "the point at which several edges intersect" or "the point at which the contour has a significant curvature change."

It is then possible to determine a "vector" between the initial and the final vertices (the direction of the vector can be chosen in such a way to guarantee a given direction, for example, clockwise, for each closed contour of the scene) and, accordingly, to classify the lines as "straight," "concave," or "convex."

Furthermore, curved lines can be approximated by circular arcs, which can be identified by the following attributes: i) length, ii) initial slope, and iii) final slope.

Then it is possible to associate with each line a LISP-like descriptor of the following type:

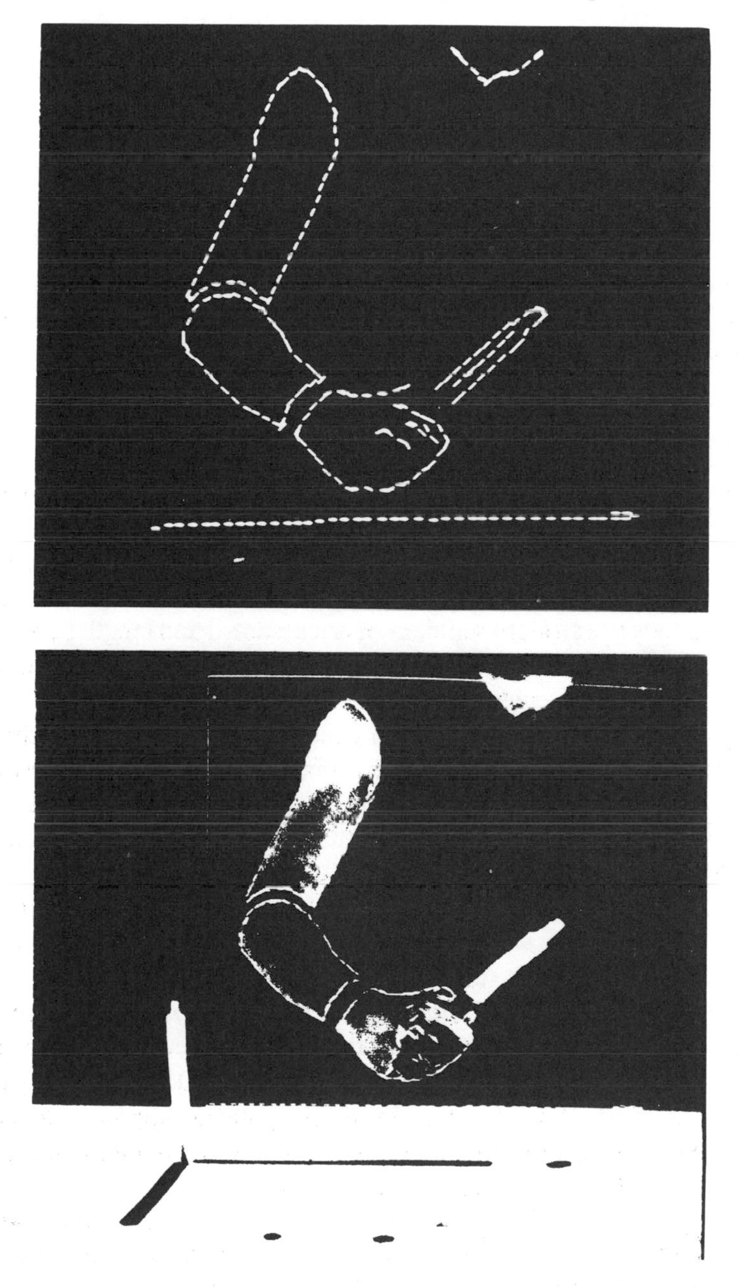

Figure 2. The image of Figure 1 is transformed into the images of Figure 2 (a and b) through feature extraction

```
(<line name> (CLASS PRIMITIVE LINE)
             (TYPE <S/C/V/R>)
             (VEXTOR <vertex-1> <vertex-2>)
             (LENGTH <length-value>)
             (INITIAL SLOPE <angular-value>)
             (FINAL SLOPE <angular-value>))
```

where S/C/V/R stand, respectively, for "straight," "convex," "concave," and "rough" (meaning non-recognizable as one of the previous three).

Recognition of Primitive Lines

Line recognition and classification can be performed using syntactic pattern recognition [8].

According to the labeling scheme, previously defined, a line can be described by a string of symbols. For example, the string "ddddd" corresponds to a straight segment of orientation d and length 5, the string "gfedcb" and the string "bcdefg" correspond, respectively, to a concave and a convex segment.

If we have grammars which can generate strings for each of the three classes (straight, concave, convex), line recognition can be performed by trying to parse a line with the three grammars: if one of the parsers succeeds, the line is classified accordingly, otherwise it is classified as "rough."

It can be easily shown that regular grammars are sufficient in the noise-free situation and, as a consequence, the parsing could be performed by three finite state automata.

Noise and feature extraction errors cause the presence of spurious symbols in the strings describing the lines. To account for noise and errors, it is necessary to use some stochastic modeling technique. On the other hand, if we want to keep the syntactic approach to pattern recognition, it is convenient to refer to a stochastic regular grammar and a stochastic finite state automaton [9].

The details of this formulation and the related algorithms are not reported here. (See [3].) However, to give an example of the stochastic-syntactic analysis performed by a finite state automaton, Fig. 3a shows the behavior of the functions $g(x) = 1\text{-}p_R(x)$ of the automata corresponding, respectively, to straight, concave, and convex lines, where $p_R(x)$ is the probability of the rejection state over the string

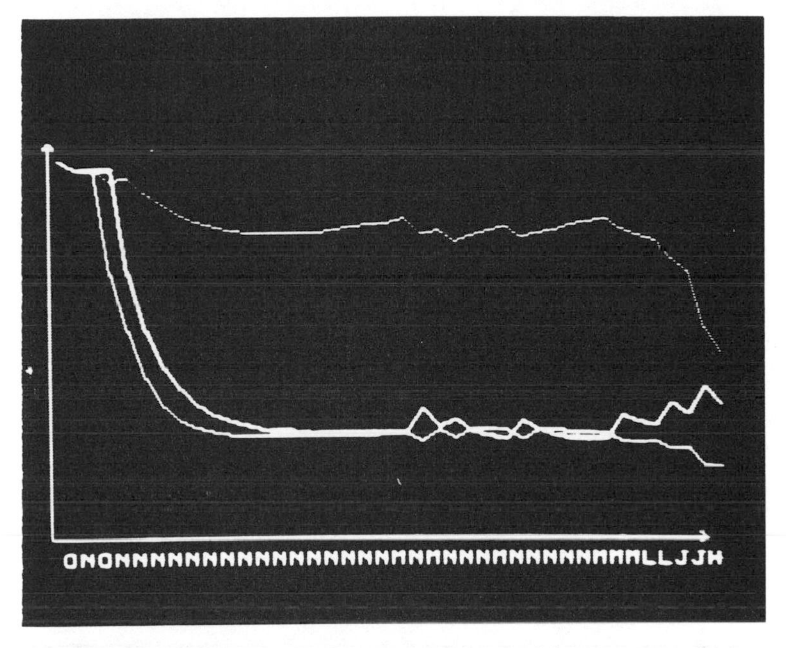

Figure 3a. The highest curve corresponds to the g (x) of the automaton which recognizes straight lines

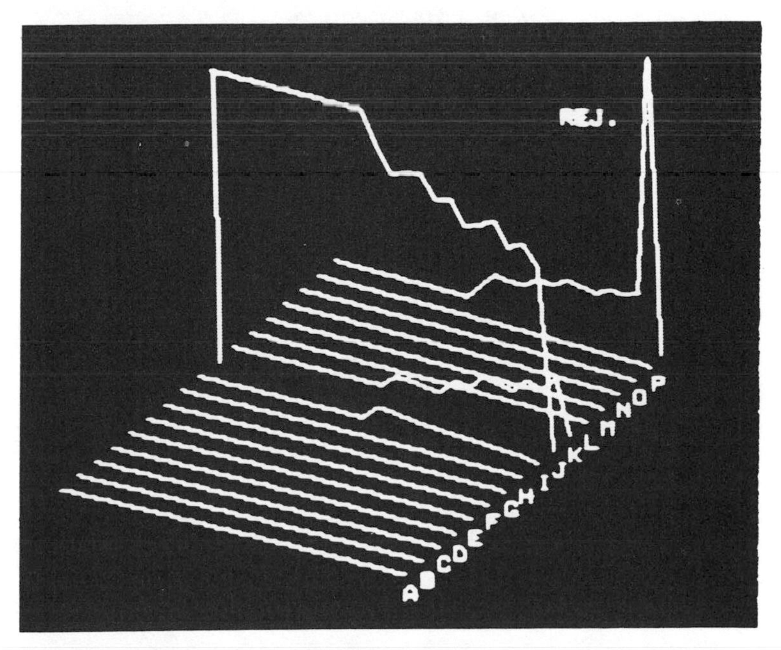

Figure 3b. Behaviour of the probability distribution of states of automaton related to straight lines during analysis of the given string

"JJJJJJJJJKIJJKJKJJKJKJK" chosen in the example. The highest curve in Fig. 3a corresponds to the g(x) of the automaton which recognizes straight lines. Fig. 3b shows the behavior of the probability distribution of the states of the automaton related to the straight lines (each state corresponds to a local orientation of the line) during the analysis of the given string.

Formal Description of a Scene

The initial analysis of a scene, at the level of edges and lines, has greatly reduced the amount of information to be processed in order to interpret the scene, by providing a data set (the set of LINE-descriptors) which is appropriate for the analysis and interpretation of the scene.

In order to outline the following levels of analysis, it is convenient to define a formal description of the scene in which we can embed useful concepts and methods:

<scene> ::= {<object> <global operator> }* <object>
<object> ::= {<figure> <external operator>}* <figure>
(5.1) <figure> ::= {<figure> <internal operator>}* <figure>|
<primitive figure>
<primitive figure> ::= Triangle/Square/ . . .

The * symbol (also known as Kleene operator) means that the string inside the brackets can be repeated any number of times, including zero.

In other words, the scene can be viewed as a set of relations among objects which involve "global operators" such as "at the right of" (we shall not further consider this level of analysis), where each object is a figure or a composition of figures and each figure is a closed contour of lines.

External operators express the contiguity among closed contours as determined by the fact that two contours have some lines in common.

Internal operators are used to segment the figures into simpler figures by introducing virtual lines until the resulting figures can be considered "primitive."

Finally, "primitive figures" are defined from the properties of the lines of their contour.

Description of Objects

Each object of the scene can be associated to a LISP-like descriptor of the following type:

(<object name> (CLASS OBJECT)
(COMPOSITION <composition list>)
(CLASSIFICATION <classification list>)
(POSITIONAL ATTRIBUTES <list of values>)
(SIZE-ATTRIBUTES <list of values>))

The COMPOSITION-attribute of the object-descriptors is used to store the list of closed contours which make up the object, together with the relations among them.

In other words, the COMPOSITION-attribute is a list with the form specified in (5.1), which contains figure names and operator descriptors.

The COMPOSITION-attribute is the first to be created in the process of scene analysis. The other attributes are generated afterwards, once the classification of the object is complete.

External Operators

External operators express the fact that two figures have a line in common, completely or in part, or one contains the other. Furthermore, if a line or part of a line is in common, it is convenient to make explicit whether one figure is inside the other or not.

Accordingly, we have defined 5 operators which are shown in Table I. The external operators are used to compose figures according to the following syntax:

(<figure name> (<operator type><operator attributes>) <figure name>)

Lists of this type can be stored in the COMPOSITION-attribute of the descriptor of an object.

Table I
External Operators

Operator	γ	δ	$\tilde{\gamma}$	$\tilde{\delta}$	ε
Object	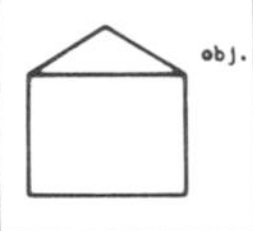 obj.	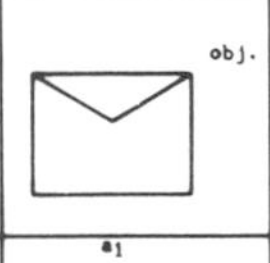 obj.	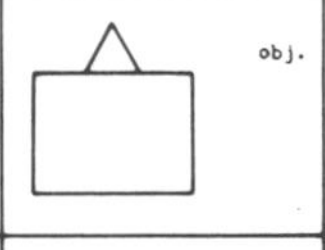 obj.	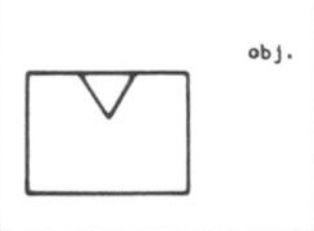 obj.	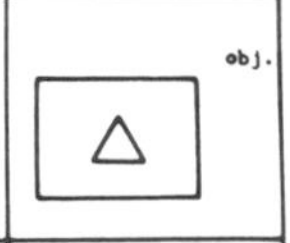 obj.
Figures	a_2 a_3 A a_1 b_1 b_4 b_2 B b_3	a_1 a_3 a_2 A b_1 b_4 b_2 B b_3	a_2 a_3 A a_1 b_1 b_4 b_2 B b_3	a_1 a_3 a_2 A b_1 b_4 b_2 B b_3	a_2 a_3 A a_1 b_1 b_4 b_2 B b_3
Composition List	Obj. COMPOSITION = $(A(\gamma\ a_1 b_1)B)$	Obj. COMPOSITION = $(A(\delta\ a_1 b_1)B)$	Obj. COMPOSITION = $(A(\tilde{\gamma}(a_1.1)(b_1.1/3))B)$	Obj. COMPOSITION = $(A(\tilde{\delta}(a_1.1)(b_1.1/3))B)$	Obj. COMPOSITION = $(A(\varepsilon)B)$

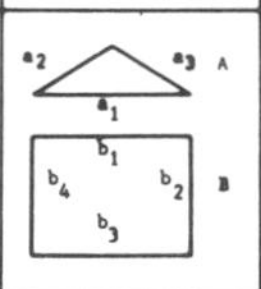

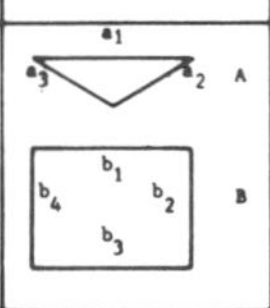

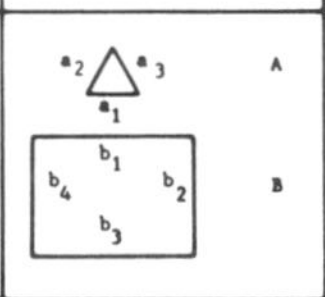

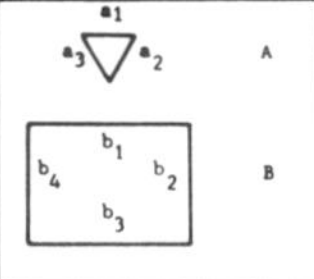

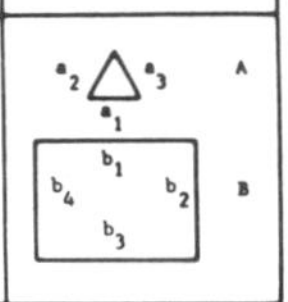

Figures

After the analysis at the level of "lines," an "object" is expressed as a set of "closed contours" related by "external operators."

The following logical step in the process of making explicit the structure of the scene consists of describing the figures in terms of their properties but, since the number of possible contours is infinite, it is convenient to rely on the definition of a small set of "primitive figures" and on a small number of operators which allow us to "segment" a closed contour into a structure of primitive figures. We shall call these operators "internal operators" (as opposed to the external operators defined in the previous section) because they introduce virtual lines for the segmentation.

Primitive Figures

Primitive figures are closed contours which exhibit certain properties, that is, specific relations among the constituent lines. For example, standard geometrical figures (such as squares, triangles, truncated circles, etc.) can be defined using relations of the following type:

- angles between consecutive lines
- parallelism between two lines
- equality between the attributes of two lines

In particular, the contour of a "square" has the following properties (1_1,

l_2, l_3, l_4 are the lines of the contour; a_1, a_2, a_3, a_4 are the angles between them; "type" and "length" are attributes of each line descriptor):

(6.1)
p1. $a_1 = a_2 = a_3 = a_4 = pi/2$
p2. l_1.type=l_2.type=l_3.type=l_4.type="straight"
p3. l_1 "is parallel to" l_3
p4. l_2 "is parallel to" l_4
p5. l_1.length=l_2.length=l_3.length=l_4.length

Furthermore, it is necessary to associate to each primitive figure an intrinsic frame of reference in order to account for different orientations, and to establish a sequencing convention of the lines.

For example, let us choose the following convention:

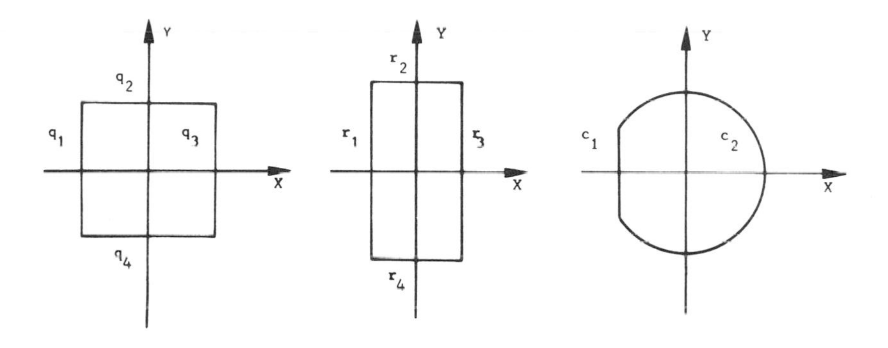

Figure 4. Primitive figures with y and *x* axes

"The intrinsic reference system of a primitive figure is a cartesian frame for which:

(i) the origin lies in the center of gravity of the figure,
(ii) the y-axis is parallel to and stays at the right of the longest straight line,
(iii) the first line of the contour is chosen as the line which intersects the negative side of the x-axis, and the rest of the contour follows the clockwise direction."

As a consequence, it is possible to define a LISP-like descriptor for primitive figures with the following structure:

```
        (<figure name>  (CLASS PRIMITIVE-FIGURE)
                        (TYPE<square/rectangle/ . . .>)
(6.2)                   (POSITION <x-coord><y-coord>)
                        (ORIENTATION <angular value>)
                        (SCALE-FACTOR <s1,s2, . . . ,>))
```

where the position and orientation attributes relate the intrinsic frame to a frame fixed in the environment.

The recognition of primitive figures can be performed by procedures which analyze the representative string of a given contour, verify the defining properties of the figure (for example, the procedure which tries to recognize squares will use properties (6.1)) and, eventually, generate descriptors of the type (6.2).

For real images it is quite clear that the verification of properties will always be associated with a degree of uncertainty, and it seems convenient to model such situations by interpreting the properties of a primitive figure as "fuzzy relations." In other words, each relation will be verified by the procedure with a "degree of fuzziness" (measured by a value in the interval (0,1)) and the global degree of fuzziness of the recognition will be computed as some weighted sum of the individual degrees of fuzziness [10].

Description of Figures

Each figure of the scene can be associated to a LISP-like descriptor of the following type:

```
(<figure name>  (CLASS FIGURE)
                (ANCESTOR <object name or figure name>)
                (CONTOUR <contour list>)
                (COMPOSITION <composition list>)
                (CLASSIFICATION <classification list>)
                (POSITIONAL-ATTRIBUTES <list of values>)
                (SIZE-ATTRIBUTES <list of values>))
```

The ANCESTOR-attribute is a pointer to an object or to a more complex figure of which the given figure is an element. The CONTOUR-attribute is used to store the list of line descriptors of the figure. The COMPOSITION-attribute is built in in the process of segmentation of the figure by recognition procedures. This attribute stores alternative decompositions of the figure in simpler figures by means of external operators.

Internal Operators

Internal operators allow us to segment a figure into simpler figures.

The decomposition of a figure into two simpler figures corresponds to tracing virtual lines between significant points of the contour of the figure. When this is done, the relationship between the two resulting figures can be expressed by means of operators (internal operators) quite similar to the external operators previously defined.

Table II
Internal Operators

Operator	ζ	η	$\tilde{\zeta}$	$\tilde{\eta}$
Object	obj.	obj.	obj.	obj.
Figures	a_2 a_3 a_1 A; b_1 b_4 b_2 b_3 B	a_1 a_3 a_2 A; b_1 b_4 b_3 b_2 B	a_2 a_3 a_1 A; b_1 b_4 b_3 b_2 B	a_1 a_3 a_2 A; b_1 b_4 b_3 b_2 B
Composition List	Obj. COMPOSITION = $(A(\zeta a_1 b_1)B)$	Obj. COMPOSITION = $(A(\eta a_1 b_1)B)$	Obj. COMPOSITION = $(A(\tilde{\zeta}(a_1.1)(b_1.1/3))B)$	Obj. COMPOSITION = $(A(\tilde{\eta}(a_1.1)(b_1.1/3))B)$

The use of the internal operators is shown in Table II.

The COMPOSITION-attribute of each figure descriptor will contain lists of composing sub-figures, related by internal operators. An important point is that each figure can be decomposed in different ways which may lead, eventually, to different classifications. The choice among different classifications can be made on the basis of the degree of fuzziness or the degree of complexity of each possible classification.

A Computational Model: The Pandemonium

The goal of a scene analysis system is to reach a description of the scene (in terms of explicit relations among primitive figures), which allows the computation of any information relevant for manipulation or other robot tasks.

The whole process can be represented, using the terminology suggested by Lindsay & Norman [11] for describing visual perception, as a "pandemonium," where a large set of "demons" detect the presence of conditions or relations in the present description of the scene and, consequently, trigger the intervention of other demons, generating other layers of description or modifying their attributes.

We have seen that relations among contours and, then among figures, can be described by means of external operators. Therefore, we need demons which correspond to external operators and that are activated by demons which verify relations among the figures, such as inclusion or line contiguity.

The process can be sketched in the following way:

(i) demons which search for inclusion or common lines receive as an input the set of contours, and give as an output pairs of contours with the specification of their relation;
(ii) the output of the previous demons are combined in order to activate demons corresponding to external operators which produce strings in which relations are translated in terms of figures (determined by contours) and external operators. Strings can be combined by observing the following rule:
(A<ext op>B), (B<ext op>C) = > (A<ext op> B <ext op> C)
After all possible combinations are made, each resulting string identifies an object and a descriptor is created for it.

The analysis continues by trying to fill all the attributes of the descriptor. As a consequence, in order to classify the object we need to interpret and classify all the figures that the contours determine.

A descriptor is then created for each figure and the analysis is carried on by trying to fill the attributes of the figure descriptors.

This process leads to:

(i) figure interpretation, that is, decomposition of a figure into simpler figures, related by internal operators, down to primitive figures, and
(ii) figure classification, which means parsing the strings obtained during the figure decomposition process by the use of grammars which correspond to particular classes of figures.

A new descriptor is assigned to each new figure created in the decomposition process and each such figure requires the same analysis, leading to recursive procedures.

Figure decomposition is carried out by three sets of demons:

(i) feature demons;
(ii) decomposition demons;
(iii) primitive figures classification demons.

Feature demons look for features which can suggest decomposition, such as:

(i) flexes;
(ii) concave lines;
(iii) convex lines;
(iv) concave angles.

Feature demons receive the contours of the figures as input and output of the sought features.

A decomposition demon is activated by a combination of the outputs of some feature demons and gives as an output a decomposition of the figure into two new figures. The decomposition is expressed by a string which contains the names of the two figures, together with an internal operator, and is written in the COMPOSITION-attribute of the figure. At the same time, a descriptor is created for each figure and a new step of recursion is taken.

When feature demons give a negative result (no feature is found), the figure is analyzed by primitive figure classifiers, which look for relations among contour lines. These classifier demons may succeed (and this stops the process) or they may recognize contour parts which are characteristic of a given primitive figure. In the case of partial recognition, the combination of the outputs of different primitive figure classifiers again activate decomposition demons, and the process goes on until no more decomposition is found (for example, the contour of a "house-figure" determines the partial recognition of the "roof" by a "triangle classifier" and of the "house body" by a "rectangle classifier," allowing a decomposition demon to segment the contour into two new sub-figures).

When the decomposition process is terminated, the produced strings are parsed by classification demons, each of which has knowledge of the grammar which generates a certain class of figures.

These classes are fuzzy classes. A figure can then be assigned to different classes, and for each class a membership value is computed. The possible classifications are written in the CLASSIFICATION-attribute of the figure descriptor.

The classification of a given figure is used in the classification process of the ancestor figure from which it is derived. In the same way, an object is classified on the basis of a classification of the figures which make it.

References

1. Chomsky, N., Aspects of a theory of syntax, Cambridge: MIT Press, 1965.
2. Parisi, D., Il linguaggio come processo cognitivo, Torino: Boringhieri, 1972.
3. Yoda, H., A new attempt of selecting objects using a hand-eye system, *Hitachi Review* 22, 362-365, 1973.
4. De Coulon, F., Kammenos, P., Polar coding of planar objects in industrial robot vision, *N.T.*, 10, 663-670, 1977.
5. Marr, D., Early processing of visual information, *Phil. Trans. R. Soc.* London, 275, 483–524, 1976.
6. Hueckel, M.H., An operator which locates edges in digitized pictures. Stanford U. *Lab. Art. Intel.*, AIM-105, 1969.
7. Carrosio, C., Sacchi, E., Viano, G., Robotic vision: an implementation of the Hueckel operator for edge detection, Genoa Un.E.E. Dept., Tech. Rep., 1980.
8. Gaglio, S., Marino, G., Morasso, P., Tagliasco, V., A linguistic approach to the measurement of 3-D motion of kinematic chains, Proceed. 10th ISIR-5th CIRT. Milan, March 5–7, 1980.
9. Fu, K.S., Syntactic methods in pattern recognition, London: Academic Press, 1974.
10. Zadeh, L.A., Fuzzy sets, *Inf. Control,* 8, 338-353, 1965.
11. Lindsay, P.H., Norman, D.A., Human information processing, London: Academic Press, 1977.

Part II:
Applications

Chapter 5

Recognition of Overlapping Workpieces by Model-Directed Construction of Object Contours

W. HÄTTICH

Fraunhofer-Institut für Informations- und Datenverarbeitung, West Germany

Abstract An image analysis method for the recognition and determination of the position of overlapping workpieces is described and tested by computer simulation. The recognition is based on a comparison of contour elements of an image with a reference contour which corresponds to a geometrical model of a workpiece and is well adapted to object data of a given image. Contours are described in terms of straight lines. Comparison is done by iteratively constructing a well adapted reference contour.

Introduction

The methods for the recognition of overlapping workpieces are based on the recognition of the shape of an object. A direct comparison of the shape as it is done, for example, by correlating pictures pixel by pixel, is very time-consuming because of the various degrees of freedom of the object's position. Therefore only the contour lines of an object are usually analyzed. Visible parts of a contour line are compared with reference contour lines which correspond to a model which may be regarded as an abstract description of the contour lines of an object. Various methods being developed differ in the comparison strategy and in the type of the models.

There are two basically different comparison strategies. In the first strategy visible parts of given contour lines are separately interpreted as parts of a reference contour which are in accordance with a geometrical model of an object. The recognition is achieved when the single interpretations are consistent. In the second strategy a reference contour is constructed using visible parts of contour lines as construction elements and using production rules as a generative model of an object. Recognition can be performed if a reference object is constructed fairly completely. Most of the methods quoted in the literature are based on the first strategy (1–3). Here, the second strategy is applied. The advantage of the second strategy is that all parts of the reference contour can be constructed successively in a prescribed manner and also that the sequence of the single comparison steps does not depend on the contour parts of the scene which may be accidentally visible or not. Additional advantageous features of the second strategy are discussed in (4).

The models reported in the literature differ in the type of description elements and in the nature of the description. According to the comparison strategy most systems rely on a fixed geometrical model, using descriptive features for the definition of the contour of an object. Here the model is defined by a set of production rules. By applying these rules reference contours can be constructed. At present two types of model are under development. One model uses corners, the other straight lines, for defining a reference contour. It is planned to combine both models in order to be able to construct any shape of a workpiece by the most suitable elements—either by piecewise linear approximation of curves or by the positioning of corners. The system using a model of corners is described in (4) and practical considerations of that system are discussed in (5). In the following the system using straight lines is presented.

The Scene

The recognition system is outlined for scenes having a complexity depicted in Fig. 1a and Fig 1b. Scenes with a comparable complexity are to be found when workpieces are isolated by simple strip-off mechanisms. Then, up to four or five workpieces are in the scene. Apart from the overlapping of the workpieces, perspective deformations and bad contrast situations occur. Perspective deformations, however, are limited as long as the workpieces are relatively flat. For testing the system screws and bent metal parts have been used.

The Model

The model of a workpiece is an abstract description of its shape. Here the workpiece is characterized by an arrangement of straight lines. The model is an ideal arrangement of straight lines which corresponds to the contour lines of an object.

The basic element of the model is a segment. A segment is a straight line which is assumed to be a part of an imaginary ideal reference object. The position of a segment is given by the coordinate values of its start

Figure 1. Typical scenes of overlapping workpieces.
(a) Scene with bent metal parts. (b) Scene with screws

and end point or, alternatively, by the coordinate values of its start point and its length and orientation.

A model is represented by a sequence of segments. Beginning with a start segment a complete ideal reference object can be constructed by a successive determination of the position of segments. In order to characterize the contour of a given workpiece the relations between segments must be specified accordingly. The consecutive determination of the position of lines has the advantage that arrangements may be defined which depend on relative values of length, orientation, and position between single lines and not on absolute values. The relative position may vary within a given range. The range of variation can be selected for each segment separately.

Fig. 2 shows arrangements of segments which characterize the workpieces depicted in Fig. 1. The segments of a model are marked by thick lines. Dashed lines show contour parts of the objects which are not taken

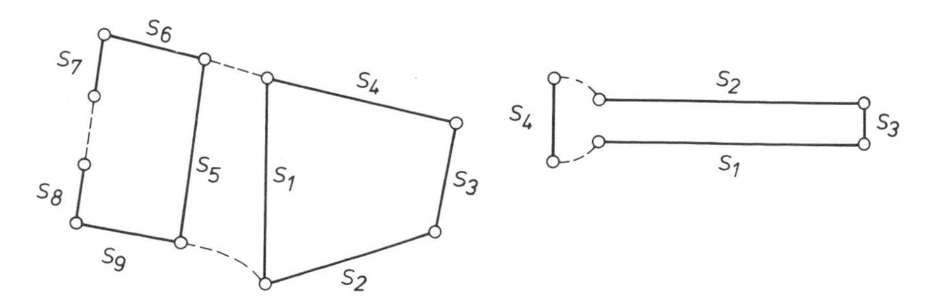

Figure 2. Models of workpieces.
(a) Model of a bent metal part. (b) Model of a screw

into consideration by the model. The number at each segment shows its order in the sequence of segments. Obviously, adjacent segments of a sequence need not touch each other.

The Scene Representation

In accordance with the model, the object contours of a scene are approximated by straight lines and a scene with workpieces is represented by such approximating lines.

The lines used for approximating the object contours are extracted by a two-step procedure. The two procedures are a contrast enhancement procedure using nonlinear local operators, and a line finding procedure using linear regression.

Both procedures are described in (4, 5). Here only the results are presented. Fig. 3b shows lines extracted from a contrast enhanced scene depicted in Fig. 3a. Due to light reflections and contrast enhancement the contour lines of the objects are sometimes to be seen as double lines. Every line in the scene is numbered. The line number, the positions of the start point and end point, and the length and the orientation of each line are stored in a table, called a scene table. For the actual recognition of workpieces only the scene table, not a picture of the scene, is used. This implies a considerable reduction of input data.

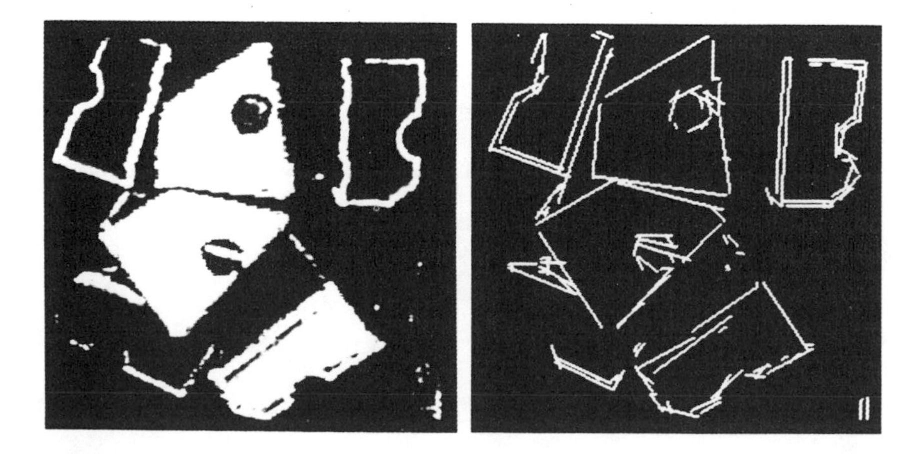

Figure 3. Approximation of object contours by straight lines.
(a) An example for a scene (contrast enhanced). (b) Approximating lines

Production Rules for Reference Objects

The reference object is an arrangement of selected lines of the scene table which is in accordance with the model of an object.

The construction of a reference object is directed by a state transition diagram. In this diagram each segment of the model is represented by a state. A state is described by two sets of coordinate values: the "real position data," and the "ideal position data," of a segment. The position data are given either by coordinate values of the segment's two end points or, alternatively, by coordinate values of one point, length, and orientation. Every state except the final state has pointers to following states and pointers to reference states. Initial pointers mark one or more start states. Several start states exist when the construction of a reference object may begin at different segments of a model or when more than one model exists. Several models exist when different objects are in the scene or when an object has different shapes for different perspective views.

Traversing the state transition diagram is possible by means of production rules. The production rules consist of two types of operation.

The first operation is to traverse to a new state. The geometric interpretation of that operation is to place a line of the scene table in order to begin a new segment.

The second operation is to remain in a state. The geometric interpretation of that operation is to place a line of the scene table in order to complete a segment.

A simplified version of a state transition diagram of the model depicted in Fig. 2b is to be seen in Fig. 4. The pointers to follower states are marked by arrows to the corresponding states. The pointers to reference states are given by a list below each state. The pictorial interpretation of the description parameters of a state is given below the state transition diagram.

For each operation a sequence of sub-operations is carried out. Each of these sub-operations can be modified and adapted to a given workpiece by a set of parameters which are associated with every transition in the state transition diagram.

The single sub-operations of the model are independent of each other. Therefore, the model can easily be modified by redefining certain sub-operations. At present the model is so flexible that an adaption to new workpieces is possible by simple changes of parameters. Suitable parameters are determined by experiments during the training phase.

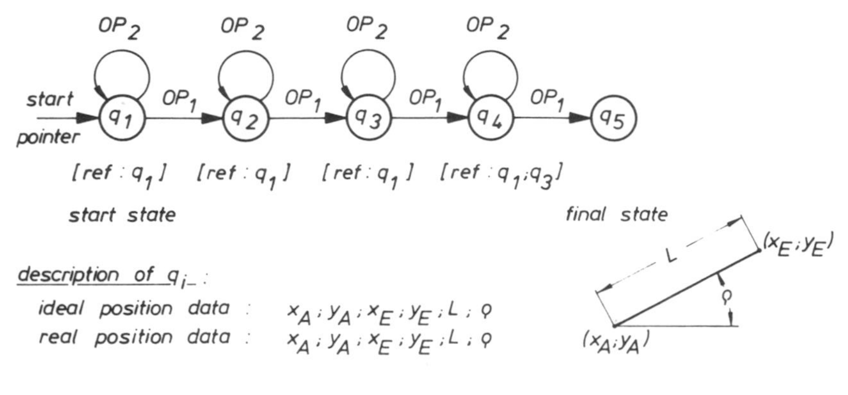

Figure 4. Simplified version of a state transition diagram of the model of Figure 2b

In the following, the sub-operations are described and demonstrated by the construction of a regular triangle.

First Operation: Starting a New Segment

The sub-operations of the first operation are as follows.

Selection of a New State

The task of this sub-operation is to check which states are follower states of a current state. If there is no follower state the current state is a final state and the construction process is finished. If there are more than one follower state different sequences of segments representing alternative models can be constructed. For each follower state all sub-operations are repeated. The list of follower states is assigned to each state, when the model is defined.

Selection of Reference States

The task of this sub-operation is to select those states which represent segments of the model on which the position data of the new state depend. The selection of reference states is necessary because the length and position of a new segment should be adaptable to the size and position of previously located segments. The list of reference states is assigned to each follower state of an operation.

Determination of Reference Data

Using the position data of the selected reference states, a reference point, a reference orientation, and a reference length are computed. These reference data are used as standards when the position of a new segment is determined. The intersection point of two segments, and the start point or end point of any segment already located can be used as a reference point. The reference length and reference orientation can be chosen separately from single segments or can be determined by averaging the values of more than one segment. A specific rule for the determination of reference data can be selected by a parameter which is assigned to each follower state of an operation. Fig. 5 shows a situation where the reference point is an intersection point of two segments and the reference length and orientation are taken from a single segment.

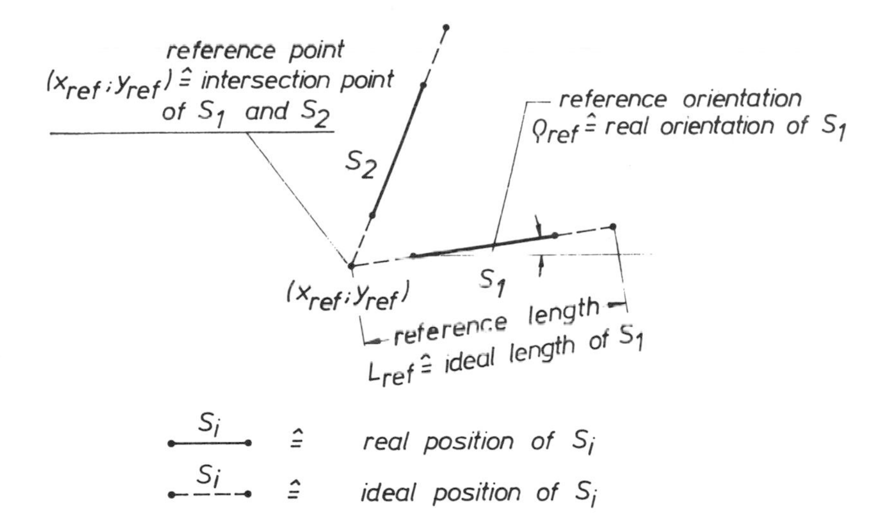

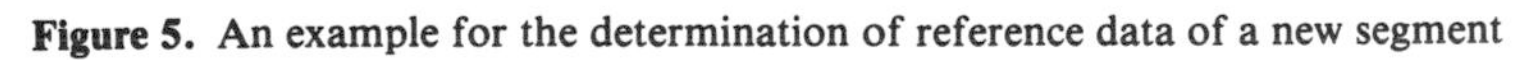

Figure 5. An example for the determination of reference data of a new segment

Determination of Ideal Position Data

Using the reference data and relative position data (which can be chosen arbitrarily when defining the model) the ideal position data of the new segment are computed. For the computation of the ideal position data the most appropriate coordinate system is used when a special relation between orientation and length of segments is requested. A cartesian coordinate system is used when simple translations of segments are required. A formula for the determination of ideal data can be selected by a parameter which is assigned to each follower state of an operation. The relative position data needed for the computation of the ideal position data are also assigned to each follower state. An example for the determination of the ideal position data of a new segment is given in Fig. 6.

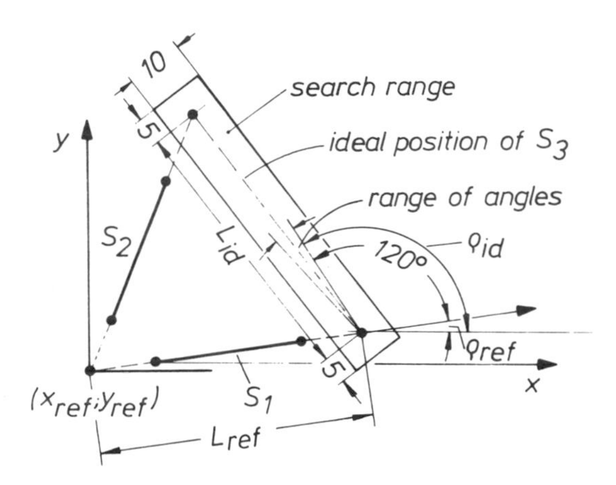

ideal start point of S_3: $(x_A; y_A) = (x_{ref}; y_{ref}) + (L_{ref} \cdot \cos\varphi_{ref}; L_{ref} \cdot \sin\varphi_{ref})$

ideal length of S_3: $L_{id} = L_{ref} + 0$

ideal orientation of S_3: $\varphi_{id} = \varphi_{ref} + 120°$

search range :

range of angle : $\pm 5°$; breadth : 10 ; additional length = 10

Figure 6. An example for the determination of ideal position and search range

Determination of a Search Range

The position of segments may vary only within a limited range. Therefore, any line which is a part of a given segment must lie within this range. A search range corresponding to the range of variation is defined by a range of angles and a stripe of points along the ideal position of a segment. A range of angles, an additional length, and the breadth of the stripe can be defined by tolerance data, which are assigned to each follower state. Fig. 6 shows an example of a search range.

Determination of Real Position Data

The task of this sub-operation is to check which lines are in the search range. Every line which lies in the search range can be regarded as a part of the new segment. If there is more than one line in the search range different realizations of a reference object can be constructed depending on which line is used for beginning the new segment. The real position data of a realization are taken from the position data of the line, used for beginning the new segment. If there is no line in the search range the ideal position data are taken as real position data storing the fact that a gap of a whole segment has been bridged. The real and ideal position data of the states of each realization are stored. In Fig. 7, an example with two lines in the search range is given. Taking each line as a part of the new segment, two new realizations are constructable.

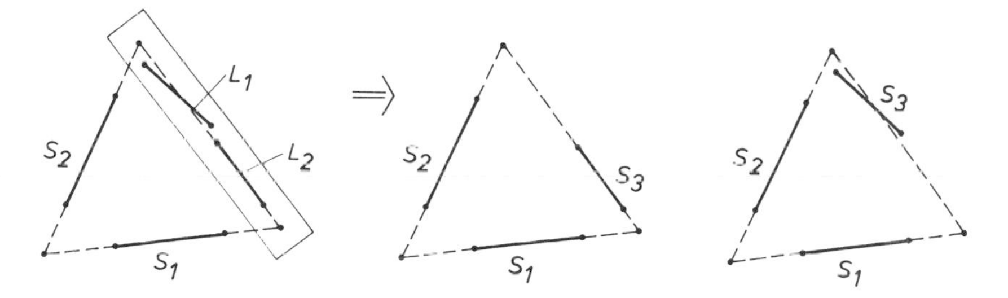

Figure 7. New realizations of a reference object when two lines are in a search range

Second Operation: Extending a Segment

Only two sub-operations are needed for the second operation, because the ideal position data of a segment are already known when applying the second operation. The sub-operations of the second operation are modified versions of the last two sub-operations of the first operation.

Modification of the Search Range

The second operation is applied when the length of the real position data is shorter than the ideal length of a segment. Therefore, any line which is appropriate for extending the line of a reference object must be located between the real line endpoint of the reference object and the ideal endpoint of the corresponding segment. The search range is modified accordingly. It is redefined by a new stripe of points along the line between the ideal endpoint of the segment and the real line endpoint of the reference object. The tolerance data defining the breadth, an additional length, and the range of angles are not modified. A modified search range of a reference object depicted in Fig. 7 is shown in Fig. 8.

Updating the Real Position Data
As in the first operation, an examination is made of which lines are in the search range, which is modified here. Every line which is located in the search range can be used for extending the line of a reference object

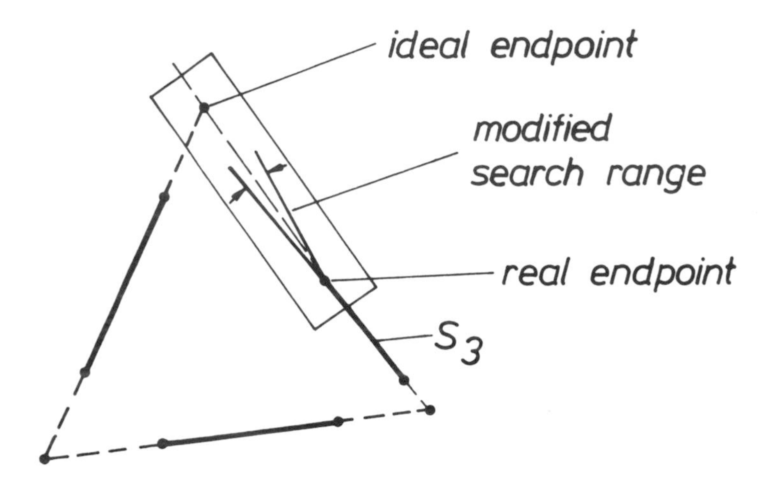

Figure 8. Modified search range for a realization depicted in Figure 7

which represents the actual segment. Again, different realizations of a reference object are constructable when more than one line are in the search range. The real position data of a new segment are updated by taking the endpoint of any line which is located in the search range as a new endpoint of a reference object. In doing this, gaps between the endpoint of the old realization and the start point of the new line are bridged. (See Fig. 9.) The total length of the gaps is stored for each realization. In the case that there is no line in the search range the whole length of the gap between the endpoint of the old realization and the ideal endpoint of a segment is stored and a further traversing operation to a new state is activated.

The Recognition Procedure

For the recognition of a workpiece the lines of the scene table are used in order to construct a reference object which fits the model. Several reference objects are constructed by applying the production rules to different lines of the scene table. Due to bad contrast situations, deformations of perspective, mutual overlap of objects, and imperfect extraction of lines, no reference object will be constructable which exactly matches the model. Therefore, the recognition process consists of constructing a reference object which is fairly complete and has a high resemblance to the ideal arrangement of the segments of a model.

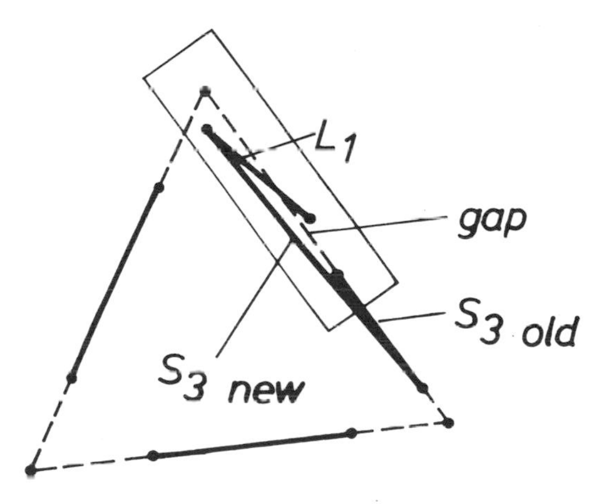

Figure 9. An example for updating the real position data of a segment

The feedback system sketched in Fig. 10 is used for iteratively constructing the reference objects (4). During each cycle of the feedback loop a sequence of four actions is performed:

Constructing of new representations of incomplete reference objects.

Comparison of the new reference objects with the model.

Storing of reference objects with a large value of resemblance to the model.

Selection of the reference object with the largest value of resemblance to the model.

The feedback structure of the recognition system can be adapted to different situations and problems:

A pure structural recognition system emerges if only those reference objects are stored which exactly match the model (6).

A gradient strategy is present if in each iteration only the best representation of a reference object is stored (7).

A branch and bound strategy results if the selection is directed by an estimate of the maximum resemblance which is attainable at a given state (4).

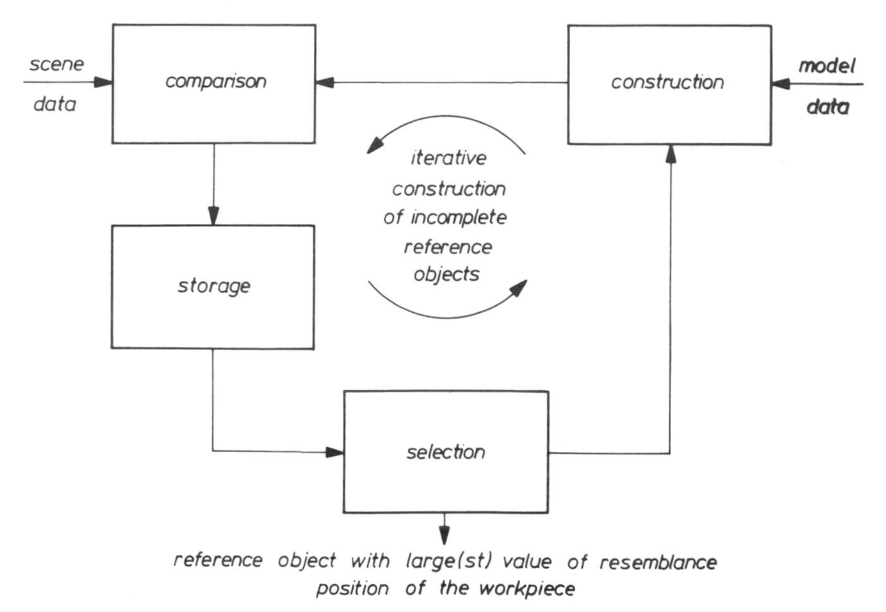

Figure 10. Feedback system for constructing reference objects

An evolution strategy is given if the selection of a reference object is random and if during the construction process only a limited number of reference objects with high resemblance survive (7).

Here a strategy similar to the evolution strategy is applied. Instead of selecting the reference object at random, a "best first" strategy is used for selecting reference objects. The actions performed in each iteration are described below.

Construction of Reference Objects

According to the production rules, reference objects are constructed using lines of the scene table, so that arrangements result which are in accordance with the model. At the beginning of the construction process every line of the scene table is interpreted as a representation of an incomplete reference object which is in its start state. The position data of the lines of the scene table are taken as real position data of the start state of the reference objects. The start state of each reference object corresponds to the start segment of the model.

Only a limited number of reference objects will be constructed. Therefore the reference objects are ordered corresponding to the real and ideal length of their start states. Only those reference objects are considered and stored of which the states have a real length almost as long as the ideal length given by a model. Reference objects with start states of shorter length are taken into the store as long as the limiting number of reference objects has not been reached.

Starting from the start state of all stored reference objects, new representations are constructed. The construction of new representations is done by traversing the state transition diagram. In order to do this the production rules described in Section 5 are applied to the lines of the scene table.

Comparison and Determination of Resemblance

Every new representation of a reference object which has been constructed by the production rules is compared with the corresponding ideal position data of its states which correspond to the segments of the model. The following two criteria are used for the definition of the resemblance.

The degree of completeness
defines how far the construction process has gone already. It is computed by summing up the ratios of the real length and the ideal length of the states which have been traversed up to the actual state of a given reference object.

The degree of distortion
takes into consideration all effects which make a complete match impossible. The degree of distortion takes into account the length of gaps when bridging missing parts of the model, deviations between real and ideal orientations of the states and additional unwanted lengths when the real length of a state exceeds the ideal one.

Each amount of the different effects can be weighted arbitarily, so that the severity of distortions which are tolerable can be influenced and adapted to the specific problem. The total resemblance is the difference value between the degree of completeness and the degree of distortion.

Storing Reference Objects

In every iteration the number of reference objects is enlarged except that the number is limited. Limiting the number of reference objects is done by storing in each iteration not more than a given number of new representations. Those new representations with the lowest degree of distortions are entered into the store. The reference objects in the store are ordered according to the value of resemblance. The size of the store is bounded. If the store is complete already, then when a new representation is to enter, the reference object with the smallest value of resemblance is canceled. When applying this strategy only reference objects are in the store which have a low degree of distortion and a large value of resemblance.

Selection of a Reference Object

For selecting a reference object in order to begin with a new iteration the "best first" strategy is used. From the stored collection of reference objects into which those objects, which have a low degree of distortion, entered, the one with the largest value of resemblance is selected. Applying this strategy, the reference objects with the largest values of resemblance are preferred. The best reference object has preference as long as its value of resemblance enlarges. When its value of resemblance is reduced, however, the object is selected next which now has the largest value of resemblance.

Two criteria are given for finishing the iterative construction process. The process finishes as soon as

The reference object with maximum resemblance has arrived at its final state, and its value of resemblance has reached the limit of acceptance. In this case a workpiece is recognized.

The reference object with maximum resemblance is below a limit of rejection. This means that by further iteration steps it cannot be expected that the value of acceptance can be reached. In this case no object is recognized.

The reference object which has the largest value of resemblance when the first criteria is fulfilled is regarded as the object which is to be recognized. The desired information about the object's position is taken from the position data of the reference object.

Results

The iterative construction of the object contours have been applied to 20 scenes with bent metal parts and screws. In no scene was the position of an object determined wrongly. In 17 scenes one or more workpieces were correctly recognized and only in 3 scenes was no object found. Fig.

11 shows two samples of such scenes. The objects which were recognized are marked. All lines of the scene table used for constructing the reference object are marked by a common number. The number shows which lines belong to the same object.

The computing effort for recognizing an object depends on the scene complexity and on the type of the model. When analyzing scenes with the screws, 10 iterations on the average are needed for the construction of a reference object which leads to a recognition. For scenes with the bent parts about 60 iterations are necessary.

For further development it is planned to combine the model with corners and lines in order to be able to enlarge the complexity both of the scenes and the objects. As a next step, objects with curved contour lines will be analyzed. Because of the flexibility of the model it is easy to change the model in order to adapt the system to new objects.

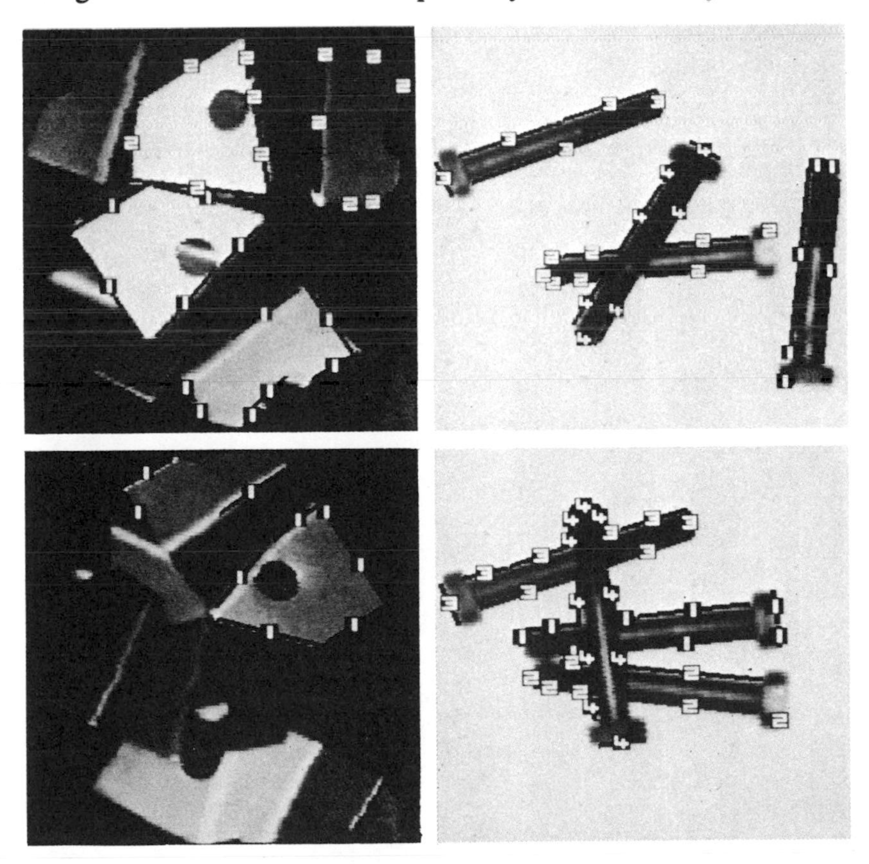

Figure 11. Recognition results for different scenes.
(a) Scenes with bent metal parts. (b) Scenes with screws

Acknowledgment

The research reported in this paper was supported by the Bundesministerium für Forschung und Technologie of the Federal Republic of Germany under contract 08 IT 5807.

References

1. W. A. Perkins: A Model-Based Vision System for Industrial Parts. *IEEE Trans. on Computers,* Vol. C-27, No. 2, February 1978, pp. 126–143.
2. B. Neumann: Interpretation of Imperfect Object Contours for Identification and Tracking, Proc. of the 4th Int. Joint Conf. on Pattern Recognition, Kyoto, Japan, 1978, pp. 691–693.
3. J. D. Dessimoz, M. Kunt, and J. M. Zurcher: Recognition and Handling of Overlapping Industrial Parts, 9th Intern. Symp. on Industrial Robots, March 1979, Washington D.C., pp. 357–366.
4. H. Tropf: Analysis-by-Synthesis Search for Semantic Segmentation, Applied to Workpiece Recognition. Proc. of the 5th Int. Conf. on Pattern Recognition, Miami, USA, December 1980, pp. 241–244.
5. H. Tropf: Analysis-by-Synthesis Search to Interpret Degraded Image Data, Robot Vision and Sensory Controls Conference Proceedings, Stratford-Upon-Avon, United Kingdom, April 1981, pp. 25–33.
6. K. S. Fu: Syntactic Methods in Pattern Recognition, Academic Press, New York, 1974.
7. J. Rechenberg: Evolutionsstrategie, Friedrich Fromann Verlag, Stuttgart-Bad Cannstatt, 1973.

Chapter 6

Simple Assembly Under Visual Control

P. SARAGA and B. M. JONES

Philips Research Laboratories, England

Abstract Until recently, industrial assembly has been performed either by fixed automation or by people. Fixed automation systems are in general fast, reliable, and appropriate to mass production, while manual assembly is slower but more adaptable to change. The purpose of "flexible automation" is to provide an alternative to these two existing methods.

Flexible machines are intended to be modular, easily programmed to do a variety of tasks, and equipped with sensors, such as TV cameras, to observe and react to changes in their environment. The machine interprets the visual information it receives in terms of its "model of the world" and instructs the mechanical actuators to make the appropriate action. In addition the results of the mechanical actions can be observed in order to improve accuracy and correct for errors.

This paper discusses some of the problems involved in constructing and using such a visually controlled machine for 3D assembly. These include manipulator control, lighting, picture acquisition, extracting 3-dimensional information from 2-dimensional pictures, machine calibration, system structure, and the use of multiple processors.

These problems are examined in the context of a particular class of assembly tasks, namely the vertical insertion of objects into fixtures. In particular an experimental system to perform a simple task is described in detail.

The Experimental System

The Task

The task chosen was the placement of three different size rings (each 2mm thick with outer diameters of 10, 18, and 25mm respectively) onto a tower consisting of three eccentric discs. Each disc is 0.2mm smaller in diameter than the internal diameter of the appropriate ring. In order to simulate a typical industrial situation in which the mounting point is difficult to see and is

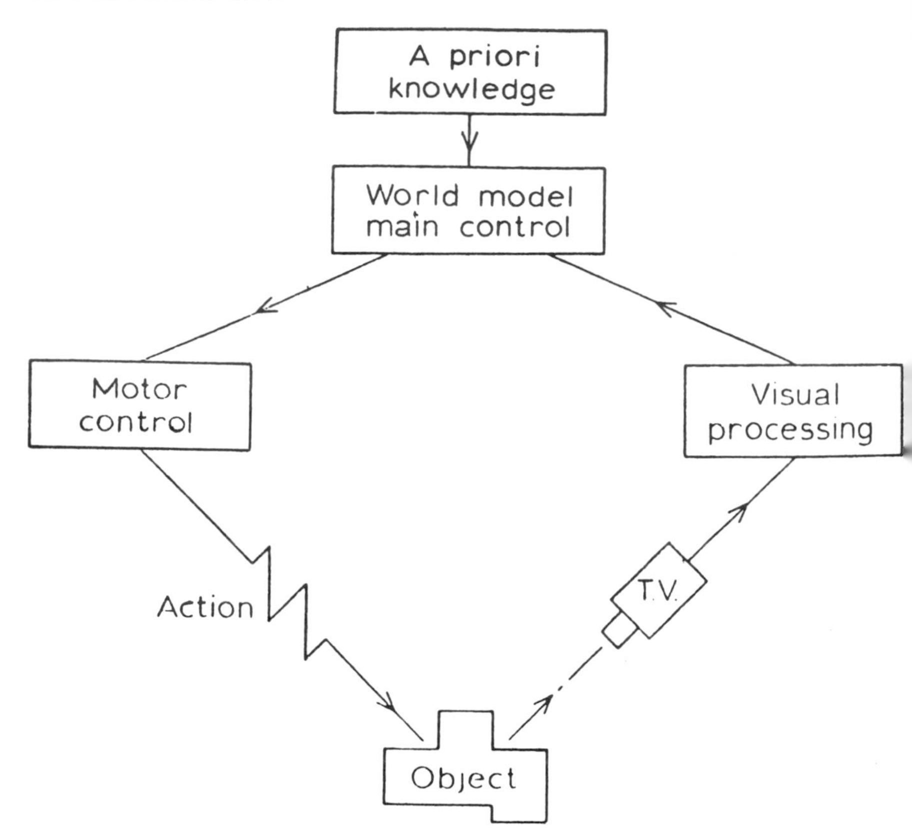

Figure 1. Conceptual structure of a visually controlled machine

part of a larger body, the three storeys of the tower are painted matt black and are mounted on a matt black base. This means that the storeys cannot be seen from above but must be located from the side. The system design should be such that the tower can be positioned anywhere within an area of approximately 70 × 70mm. The rings are assumed to lie on a horizontal surface each within an area of approximately 40 × 40mm.

Although this is not a real industrial task, it contains the key elements of a number of practical industrial problems. The task was solved by an experimental system consisting of a computer controlled manipulator equipped with TV cameras. (See Fig. 2.)

The Manipulator

The manipulator has 4 degrees of freedom of which 3 were used

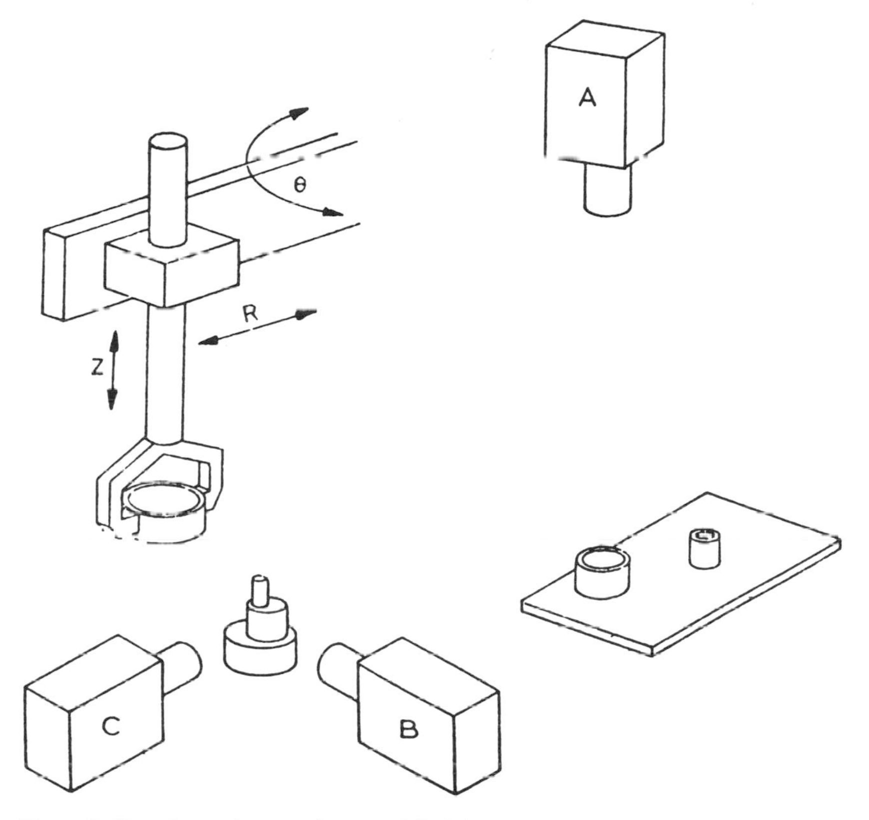

Figure 2. Experimental system for assembling rings onto a tower

for this task. Radial, rotation, and vertical (R, θ, Z) motions are provided by hydraulic actuators with integral measurement systems. Each actuator can move 100mm in 4000 steps of 25 microns each. Although the resolution is 25μ the absolute accuracy of each axis is only 0.1mm. The R and Z axes are driven directly while the θ axis has a mechanical advantage of approximately 4, giving an absolute accuracy of ±0.4mm. It can be seen that by itself this manipulator is not accurate enough to perform the task.

The fourth degree of freedom which was not needed for this task is an electrically powered wrist mounted just above the gripper. The manipulator is controlled by special purpose electronics connected to a small computer (a Philips P851).

The Vision System

The size and positions of the 3 rings are determined using a vertical

TV camera, A, as illustrated in Figure 2. The position of the appropriate storey of the tower is located by two horizontal TV cameras (B and C). Each of the two cameras sees a projected image of the tower, which has been back illuminated by light parallel to the optic axis of the TV cameras. The shape of the tower in these parallel projections is independent of the position of the tower in space and hence can be easily analyzed to give the position of the tower in the TV image. The two TV positions can then be combined by conventional triangulation to give the position of the tower in space. This method has enabled the potentially complex task of the 3D location of the tower to be divided into two simple 2D tasks.

The optical system for parallel projection, the visual processing for ring and tower location, and the combination of 2 views to locate an object in space are all described in more detail later.

The Computer System

The overall structure of the system is shown in Figure 3. It consists of two computers interconnected by slow V24 links, one link being dedicated to each direction. The TV cameras are connected to the Philips P857 by a flexible interface "TELPIN," developed as part of the same project which can handle up to 8 different channels selectable by the program. The interface can provide analog or digital information from the input channel and can select any area from the TV picture and one of 4 resolution values. The threshold value used to obtain binary pictures can also be computer controlled. The P857 performs the overall strategic control of the machine and the processing of the pictures to extract the positional information needed for the task. The smaller P851 carries out the detailed path control of the manipulator and some higher level functions involving several movements and gripper changes. Thus, once the ring has been located the P857 issues a single command to the P851 to pick up the ring and bring it over the tower. The P851 then carries out the detailed moves and grip changes independently, and informs the P857 that it has completed the action.

System Operation

The flow chart of the system operation is shown in Figure 4.

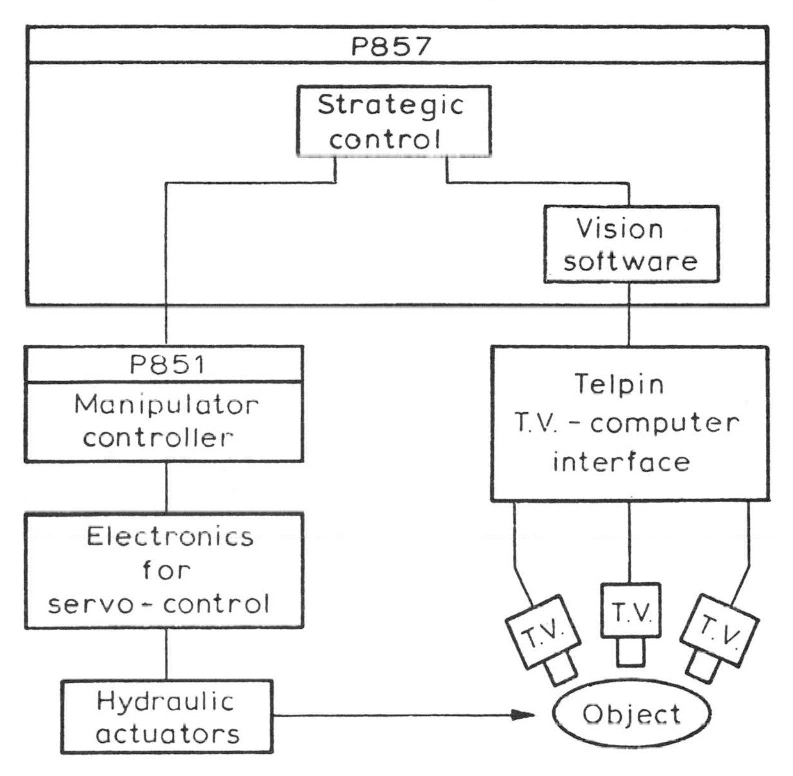

Figure 3. System structure

Once a ring has been located, and its size determined, the manipulator is instructed to pick up the ring and take it to a position high above the approximate position of the tower. This operation is under the control of the P851, and the view of the tower is not obscured by the manipulator during this period. Therefore the P857 can use the two horizontal views to determine the position in space of the appropriate storey of the tower at the same time as the mechanical operation is taking place.

If the tower storey is successfully located then the manipulator lowers the ring to a position immediately above the appropriate storey of the tower, where the relative position of ring and tower can be checked using the horizontal TV cameras. The manipulator now moves the ring until it is exactly above the correct storey. The ring is then placed on the tower.

While the P851 is controlling the final placement of the ring onto the tower, the P857 is using the vertical TV camera to locate the next ring.

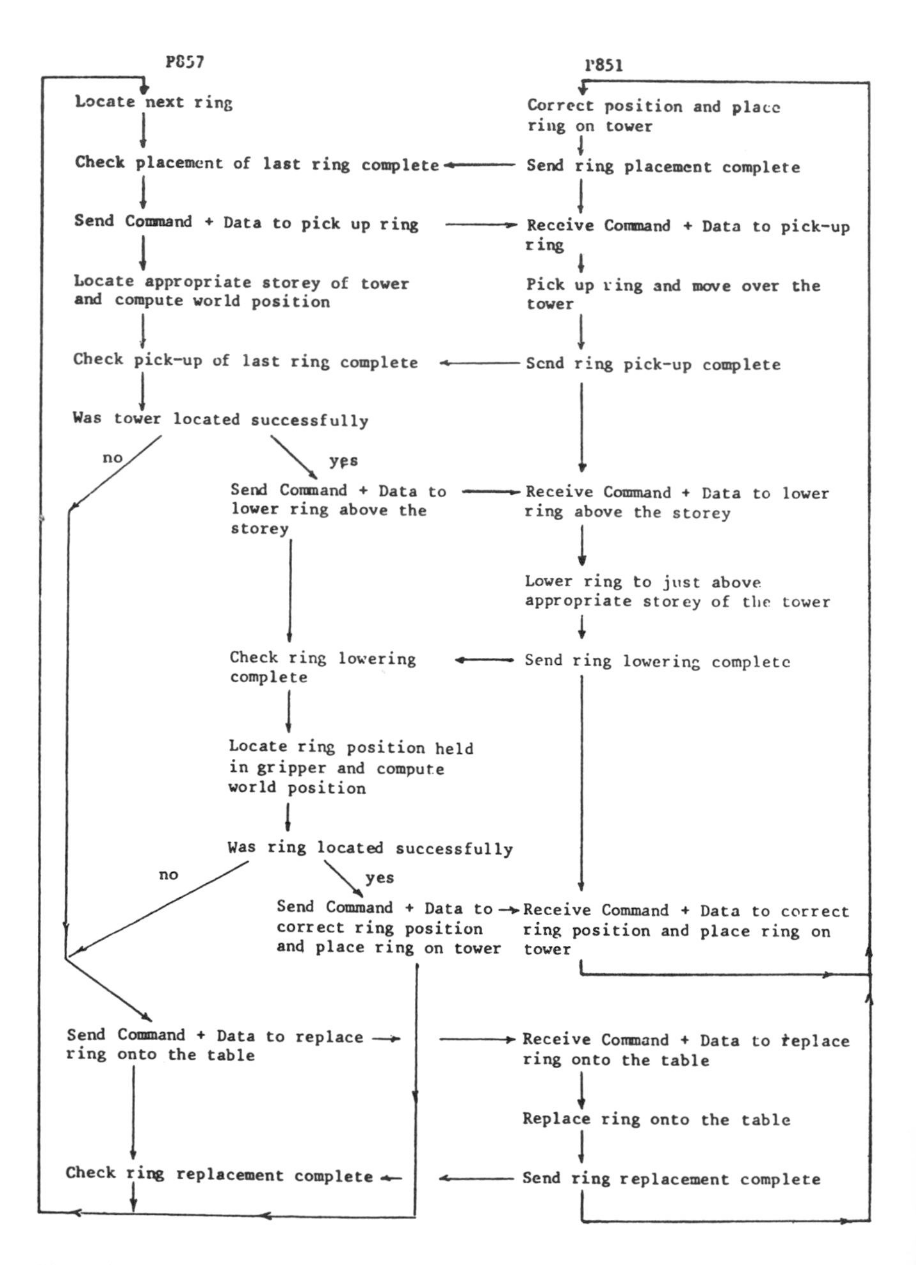

Figure 4. System operation flow chart

Picture Processing

In order that the assembly task can be carried out at a realistic rate it is necessary that the picture processing employed be fast. Thus very simple algorithms have to be employed. To illustrate this the optical processing used in the task will be described in more detail.

Ring Location

One of the three prescribed areas of the field of view (Figure 5a) of the vertical camera is sampled at a low resolution into the P857 as a 5 bit grey level image. The grey level image is thresholded and the binary image is edge traced [2], and black areas are located. If an object corresponding to one of the possible rings is not found, further attempts to detect a ring by changing the threshold are made. If a ring still cannot be found the next area is examined. Once a ring is detected its approximate center is calculated and two further regions centered on the approximate ring position are sampled at high resolution to determine the accurate X and Y coordinates of the ring centers in TV units (Figures 5b, 5c). The

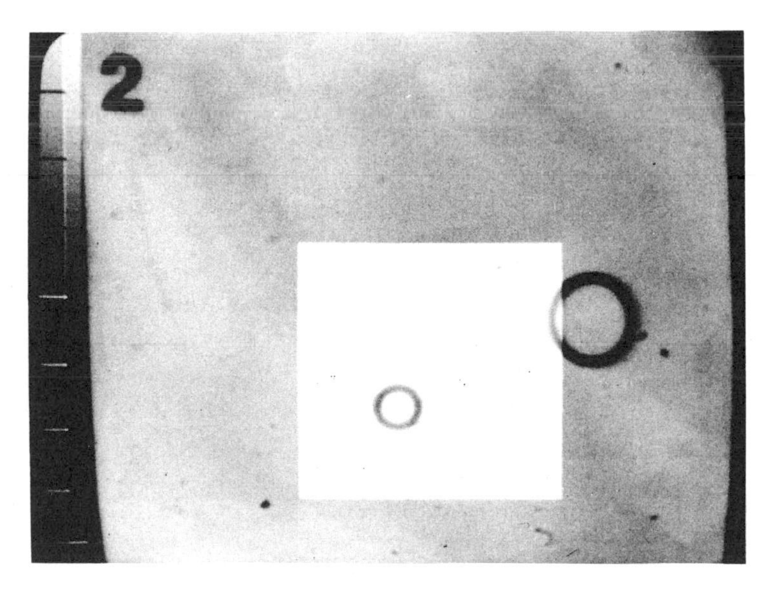

(a) Low resolution: Determination of ring size and approximate position

Figure 5. Ring location – Area 2

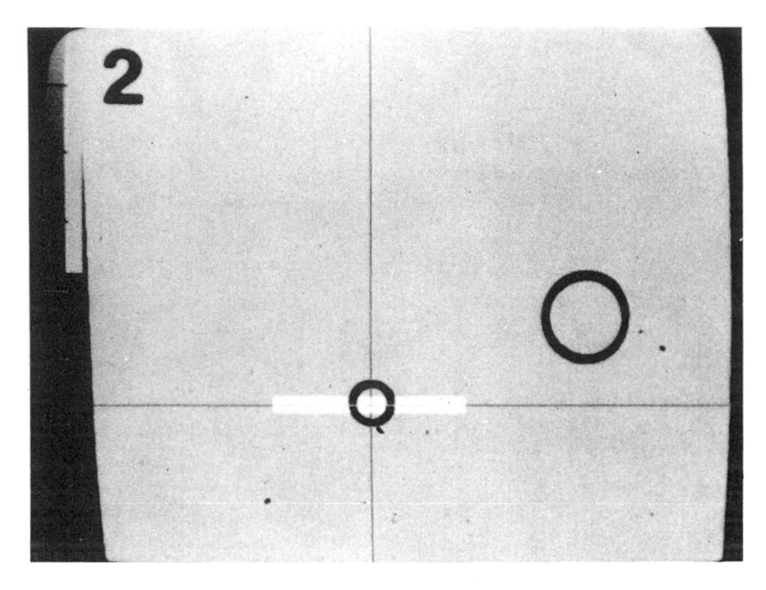

(b) High resolution: Determination of X coordinate

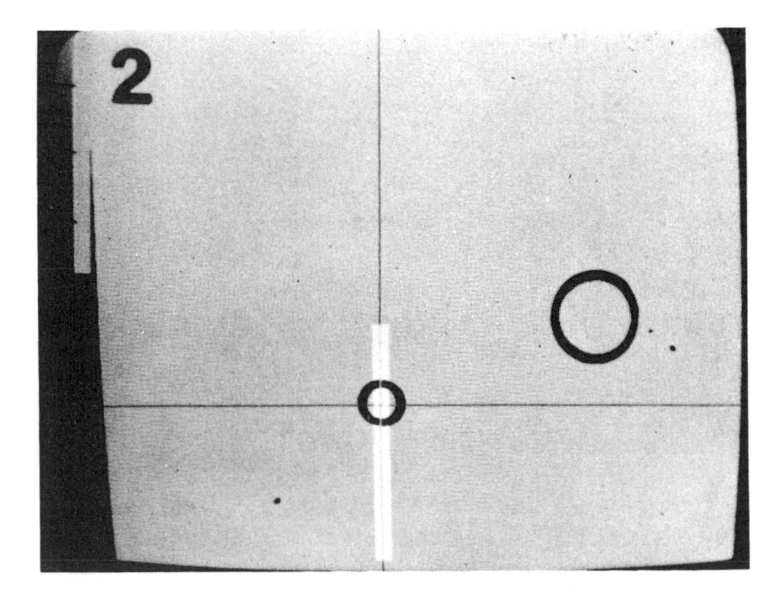

(c) High resolution: Determination of Y coordinate

Figure 5 (cont'd). Ring location – Area 2

regions are searched from both ends until a white or black edge is found and the appropriate coordinate is taken as the mean of these two edges.

The position of the ring is then converted to world coordinates and the manipulator is instructed to pick up the ring and bring it to a position approximately above the tower.

Location of Tower

The same processing is carried out on each of the two horizontal TV channels used to locate the tower. First a large coarse resolution scan (Figure 6a) is carried out to locate a black horizontal bar of the correct size for the approximate storey. Then a second smaller scan (Figure 6b) at higher resolution is taken at about the approximate center of the tower. This is used to locate the top of the storey. Small scans at high resolution are then taken at each side of the tower (Figures 6c, 6d) and the black/white boundary points found. The separation between these points is computed and only those pairs whose separation is within 5 pixels of the correct storey widths (which range from 50 to 110 pixels) are accepted. When 8 pairs have been found meeting these criteria, the mean of their center position is found and taken as the center of the storey. The mean of the error in width between the acceptable pairs and the correct storey width is used as an error function to correct either the expected width of the tower or the threshold for the next cycle of the system. Thus the system is self-correcting and can allow for drift in the TV video level. The process is repeated on the second camera. The TV coordinates are converted to a line in world space parallel to each optic axis and then the point of closest approach between the two lines is found and taken as the world coordinates of the tower. The tower may be placed in any position where all storeys can be seen by both cameras. The field of view of each camera is set to be approximately 100 × 100mm which completely covers the 70 × 70mm area in which the tower may be.

Visual Feedback

The sampling system can resolve approximately 600 × 600 points in the TV field of 10mm × 100mm giving a resolution of ≈0.17mm. The TV position of the tower is known to a rather higher accuracy since it is obtained by averaging a number of edge points. The

(a) Low resolution: Initial-search

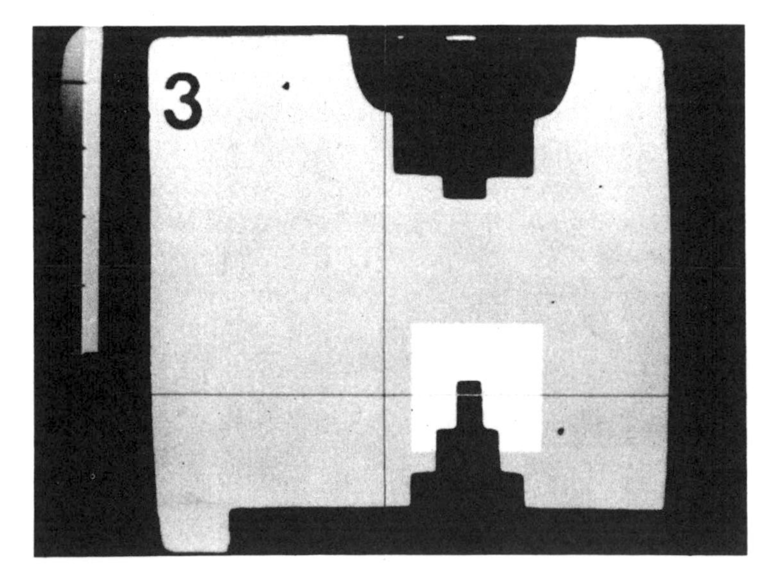

(b) Mid resolution: Determination of top of storey

Figure 6. Tower location on TV channel 3

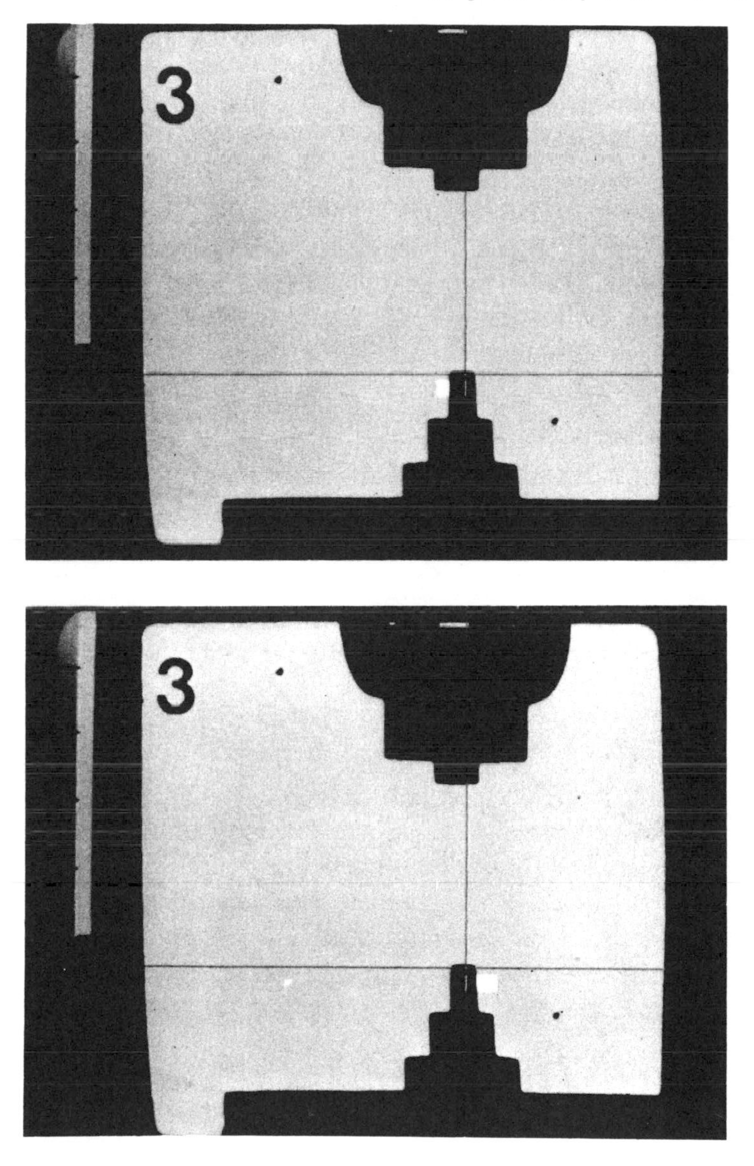

(c), (d) Final position determination at high resolution

Figure 6 (cont'd). Tower location on TV channel 3

accuracy of the world position depends both on the accuracy of the TV position and on the accuracy of calibration of the camera. Experimental results give an accuracy in the world position of roughly ±0.1mm.

However, as mentioned earlier, the clearance between the rings and the tower is only 0.1mm in radius, and while the manipulator has a repeatability of 0.025mm in R and Z and 0.1mm in θ, the non-linearity in the measuring system and rounding errors in the geometric modelling of the manipulator result in an absolute accuracy of only 0.1mm in R and Z and 0.4mm in θ. These figures, together with the inaccuracy in the location of the tower, will clearly not allow the ring to be safely put on the tower.

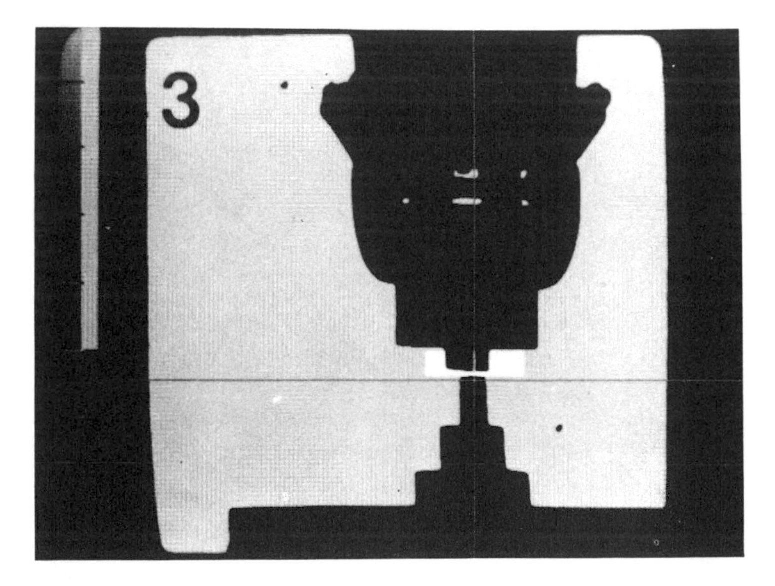

Figure 7. Checking ring position over tower on TV channel 3

This problem may be overcome by using visual feedback, that is, by lowering the ring to a position immediately above the appropriate storey of the tower and then observing the relative displacement of ring and tower in the two horizontal cameras. A narrow high resolution scan (Figure 7) is made to locate the ring projecting below the gripper. The location of the ring depends again on finding a pair of white/black and black/white vertical edges with the required separation, and the same updating mechanism is used to cope with apparent changes in ring size due to camera drift. The world displacement between ring and tower may now be calculated and the manipulator ordered to make the appropriate adjustment before placing the ring on the tower.

The use of this visual feedback has proved critical in the successful implementation of the task. It could not have been achieved with a fixed vertical camera since the manipulator obscures the critical area which has to be observed.

The visual feedback also allows the position of the ring in the gripper to be checked. If the ring is not square in the gripper then it will not be recognized and the program will instruct the manipulator to replace the ring on the table. Thus the critical region of the task is protected by a visual interlock.

The Parallel Projection Optical System

As previously mentioned, the two horizontal cameras use parallel projection optics. The system used [3] is based on that employed in the conventional tool room projector [4]. The optical arrangement is as shown in Figure 8. The object is back illuminated by a parallel beam generated by a small light source, S, at the focus of a fresnel lens, L1. A combination of a second fresnel lens (L2) and a camera lens (L3) produces a real image of the parallel projection of the object on the face of a TV camera tube, C. It will be seen that this method of viewing the object has many advantages, when applied to the determination of the position and orientation of objects.

Using this optical system, the image size is essentially independent of the position of the object. This considerably simplifies the picture analysis which can remain constant even when the object is moved nearer or further from the camera. The position of the object perpendicular to the axis is of course given directly by the position of the object in the TV image.

The optical arrangement is virtually immune to changes in ambient illumination since only light which is nearly parallel to the optic axis passes through the camera lens. The smaller the aperture of the camera lens, the higher the immunity and, since the TV camera needs relatively little light, this aperture can be made very small. This feature may be very useful in a factory environment where ambient lighting can be difficult to control. This very small aperture also gives a large depth of focus.

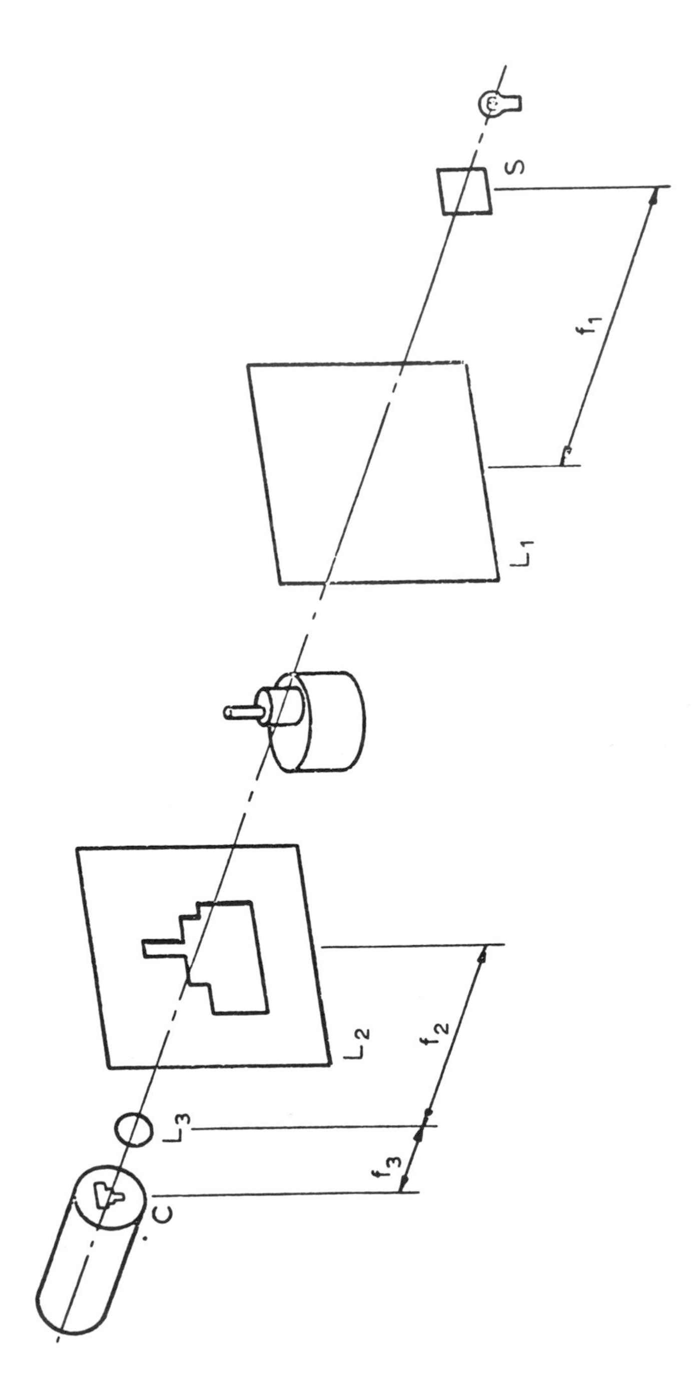

Figure 8. Optical arrangement for parallel projection

Finally, it is easy to calibrate these systems and to combine them to produce object location systems. Both these features are discussed in more detail below.

Object Location Using Parallel Optics

Because the optical systems use parallel light and are immune to ambient illumination it is generally possible to have a number of systems at various positions within a machine without the systems interfering with each other. Combination of systems is usually necessary to locate objects in space. A single axis system can only "locate" an object point as lying on a line in space parallel to the optic axis. Unless the object is otherwise constrained (e.g., the camera is vertical and the object is lying on a horizontal plane [1]) two systems are required to locate an object in space.

The location of an object in space from two views requires that the projection of a common object point be identified in each image. It is only then that the object point can be defined by the intersection of two projecting rays.

In the case of the tower, each storey is a cylinder. Such an object viewed from two calibrated horizontal cameras is illustrated in Figure 9. The common point, the center of the top surface of the cylinder, is projected as the center of the black rectangle appearing in each image. The direction of the two cameras is not critical; the more orthogonal they are, the better the accuracy of the location.

It is interesting to consider the extension of this method to other 2½D objects, by which is meant objects with vertical parallel sides whose cross section is not circularly symmetric. The projected profile seen by each of the cameras will be a rectangular black area as in Figure 9, but the width of the profile will depend on the orientation of the object to the optic axis of the camera. The problem of determining a common point in both views generally becomes more complicated, although for some regular shapes (which often occur in industry) such as a square, rectangle, or hexagon, the center of the profile still corresponds to the projection of the center of the object.

Because the object is not circularly symmetric it is probably necessary to determine the orientation as well as the position. Back lit parallel projection has an inherent 180° ambiguity since there is no difference in profile if the camera and light source are

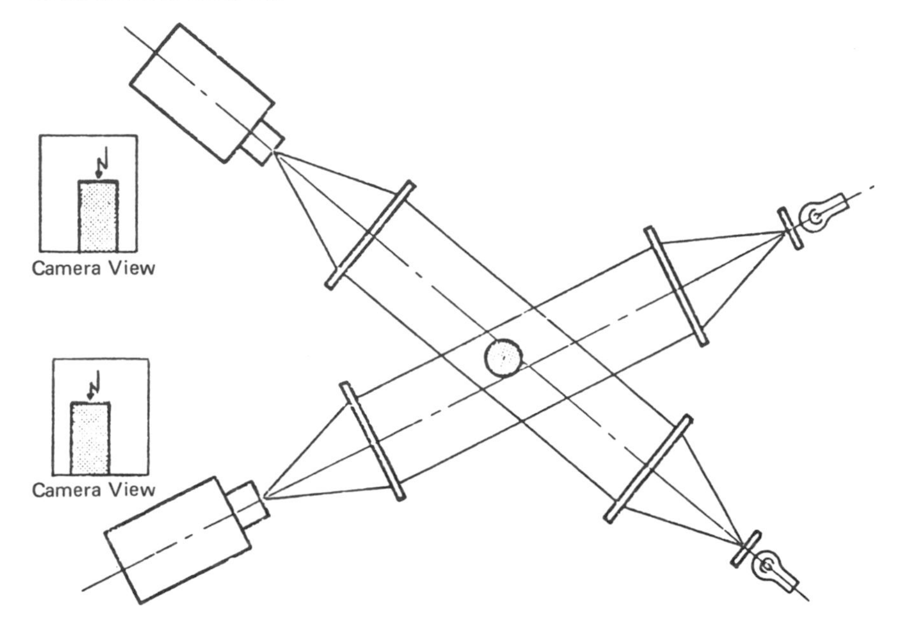

Figure 9. Location of an object from two views

interchanged. In many industrial applications this ambiguity may not be a problem, either because the assembly possesses twofold symmetry, or more generally because the crude orientation of the object is known, possibly from some previous stage of assembly. A situation of this type might occur in the final stages of assembly when the object is about to be inserted into the fixture. In this case a camera could be arranged to be at an angle where the rate of change of profile width is large.

If the approximate orientation is not known it still may be possible to remove the ambiguity using basically the same optical system. If lights are shone from the direction of the camera towards the object, then only light reflected parallel to the axis will be accepted by the camera. This results in a grey level image of the object, again of constant size, in which details of the front surface of the object can be seen. By observing this image and determining only the presence or absence of some feature it should be possible to remove the 180° ambiguity. A single measurement of the profile width will not be sufficient to determine the orientation, but using two or more cameras to view the object will in general allow the object's orientation to be determined. Alternatively, the object could be rotated by a known amount in the field of view of one camera.

An object's profile width may be determined analytically if the cross section has a simple form. Alternatively the profile may be determined experimentally and stored in a reference table in a training phase. Then, in the operating phase, the width of the profile can be measured and looked up in the table to find the corresponding value of the orientation, θ.

If the object is such that it is not easy to determine the position of a known object point in the profile, then a similar method involving the use of a table to determine both the orientation and the position may be used. For example, the position within the profile into which a particular object point, P, is projected when the projection angle is θ, can be stored in the same table as the width during the training phase. Thus in the operating phase, once θ has been found, the position of P in each projection is given by the table. The position of P in space can now be obtained by converting the TV coordinates to world coordinates and calculating the intersection of the two rays.

More complex fully 3-dimensional shapes may often contain an element of simple cross section which can be used to solve the "correspondence problem." Because the image size is constant the position of this element within the complex outline can be found by simple dead reckoning from the bottom of the object.

Calibration

In order to use a visually controlled machine it is necessary to relate the images from the various TV cameras to the movements of the manipulator and to the positions of the various parts being handled. The calibration process determines the relationship between "frames of reference" associated with each part of the overall system. Whenever the machine is modified or adjusted, the relationships change, and recalibration is required. If effective flexibility is to be achieved it is important that calibration is both simple and rapid.

It is generally most convenient to relate all the manipulator and TV frames to a single arbitrary Cartesian "world frame." Thus, if any unit is moved, only its relationship to the world is changed.

Calibration of the manipulator involves determining the relationship between the movements of the R, θ, Z actuators and the world

(a) Pointer in manipulator gripper

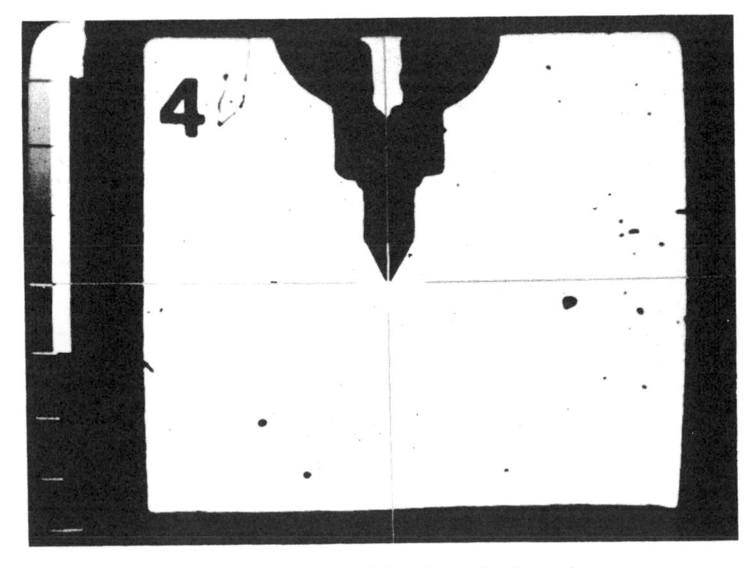

(b) TV image of 4 (a) location of pointer tip

Figure 10. Calibrating TV channel 4

frame. It can be determined from the geometrical structure of the manipulator and the "extension per step" profile of each actuator. Since the three manipulator axes being used for this task are decoupled, the geometric relationship is not complicated.

In general, calibrating a TV camera means determining where it is in 3D space and in which direction it is pointing [5]. Using the parallel optics system, it is not necessary to know where the camera is, but simply to know in which direction it is pointing relative to some "world coordinates" and also the magnification of the lens. Suppose the optical system is being used in conjunction with a manipulator whose movements are known in world coordinates. Then if the manipulator is moved through known distances in defined directions and the results of those motions observed in the TV image, the 4×4 matrix, which specifies the relationship between the TV and the world frame, may easily be calculated.

In our experimental system we place a rod with a sharp point in the jaws of the gripper (Figure 10a). Since the image of the rod remains constant in size as the manipulator moves, the point of the rod is very easy to recognize (Figure 10b), and the movement of the manipulator is easy to observe. This simple procedure allows cameras to be placed in positions and angles which suit the task. Calibration for a new position requires less than two minutes.

Distributed Processing

In the example given of positioning rings on a tower we used two processors. The division of the task between the two processors is such that only simple commands with parameters are passed to the P851, which acts as a slave of the P857. This simple division allows a degree of parallel processing and a consequent increase in overall speed of the system. There is no problem of data division in this system since all the data is stored in the P857.

It is possible to devise other organizations using more processors to allow additional processes to be carried out in parallel. For example, if a separate processor was used for each horizontal camera then the determination of the tower and ring position in each view could be carried out simultaneously.

It is necessary to consider other factors before adding processors indiscriminately. For example, the division of tasks must not require large amounts of data to be transferred between processors, nor should it require the maintenance of the same data in two processors. The distribution of processes among processors should also reflect the conceptual structure of the overall system in order to produce a modular configuration which is easy to understand and modify. Another important consideration is the relative speeds of the various parts of the system. There is obviously no point in adding processors to speed up the location of the tower if, as is already the case, the tower is located in approximately the same time as it takes the manipulator to move the ring to above the tower. There are three speed determining factors in a visually controlled machine with multiple processors. They are: the speed of the manipulator, the speed of the visual processing, and the speed of intercommunication between the processors. Changing any one of these factors can completely alter the optimal number of functions to be performed in each processor.

Conclusion

The experiments described in this paper have shown that back lit profile images using parallel light can be usefully applied to some tasks in mechanical assembly. It has been demonstrated that the advantages of this approach include: rapid size independent picture analysis allowing identity, position, and orientation to be determined by direct measurements; immunity to ambient illumination; and simple 3D object location using 2 or more views.

Although the system has only been applied to a simple example we intend to apply the same methods to a number of practical examples. The use of many processors including specialized high-speed picture processors in systems of this type is now becoming practical. It is therefore becoming important to find methods for both constructing and using these systems to give the desired qualities of speed, modularity, flexibility, and conceptual simplicity.

Acknowledgments

We would like to acknowledge the contribution of D. Paterson and A.R. Turner-Smith who, with our colleagues at Philips Research Laboratories, Eindhoven, constructed the manipulator. We would also like to thank our Group Leader, J.A. Weaver, for his help and encouragement.

References

1. Saraga, P., and Skoyles, D.R.: An experimental visually controlled pick and place machine for industry, 3rd International Joint Conference on Pattern Recognition, Coronado, California, November 1976, pp. 17-21.
2. Saraga, P., and Wavish, P.R.: Edge tracing in binary arrays, *Machine Perception of Patterns and Pictures, Institute of Physics* (Lond) Conference ser. No. 13 (1972), pp. 294-302.
3. British Patent Application No. 7942952: An object measuring arrangement December 1979.
4. Habell, K.J., and Cox, A.: *Engineering optics* (Pitman 1953, pp. 255-263).
5. Duda, R.O., and Hart, P.E., *Pattern classification and scene analysis* (Wiley Interscience, 1973).

Chapter 7

Visually Interactive Gripping of Engineering Parts from Random Orientation

C. J. PAGE and A. PUGH

Department of Production Engineering
Lanchester Polytechnic, England

Department of Electronic Engineering
University of Hull, England

Abstract This paper describes techniques for the visual analysis of complex engineering components for the purpose of subsequent mechanical handling. The experimental research rig comprises a versatile manipulator with two gripping devices together with an industrial television camera for sensory feedback. The important aspects of the work are the reliable and fast software for image analysis coupled with small computer memory requirements. Typical scene analysis times are in the range 0.5 to 3 seconds depending on the complexity of the image.

Introduction

During the early 1970's work was initiated by a small number of research groups worldwide on visually interactive robot systems. These developments were preceded by impressive research from artificial intelligence groups in establishments such as M.I.T. [1], S.R.I [2, 3], and Edinburgh [4]. One of the earliest groups active in industrial applications of visual feedback was based at the University of Nottingham, where the "SIRCH" robot became operational in 1972 [5, 6]. This robot manipulative device will be described later. About the same time, two Japanese groups reported similar interests, one group was Hitachi featuring a visually interactive machine conceptually similar to "SIRCH" [7], the other with Mitsubishi reporting experiments with an "eye-in-hand" robot applied to retrieval of motor brushes from a quasi-random presentation [8].

Over recent years the interest in visually interactive industrial robots has grown dramatically, with S.R.I. [9] and General Motors [10] reporting marketable devices that can be integrated with commercial robot manipulators. The University of Rhode Island [11] and the Swiss Federal Institute of Technology [12] have published their developments on the gripping of industrial parts from random (and overlapping) orientation. It is to this problem that the present paper is directed. The paper deals with work completed at the University of Nottingham [6, 13, 14].

The techniques described concentrate on the fundamental problems of gripping. The aim is to allow the assembly machine, under the supervision of its controlling computing system, to pick up in a precise fashion complex components of which it has no prior knowledge, regardless of whether those components are part of a heap or whether they are close to or touching adjacent parts. This facility may be an integral part of a complete assembly operation or it may form the major component of a simple sorting process. In either case, incorporating such a capability provides the machine with a substantial degree of "intelligence."

Problem Definition

The research machine used for evaluation consists of a turret assembly with three manipulators and an objective lens mounted on its periphery. The turret can be moved in three-dimensional space over a work surface by means of three mutually orthogonal, computer-controlled linear stepping-motor tables. Figure 1 shows a photograph of the machine configuration. The turret, Figure 2, can be likened to the lens assembly of a multi-objective microscope in that any of the three grippers can be indexed round into the reference position used for picking up components. A small television camera mounted over the turret assembly is used to provide the sensory feedback. A fourth station on the turret is used to mount the objective lens, which transmits an image of the scene below it up through an optical endoscope to the television camera above. The center of the machine's field of view therefore corresponds to the center of each manipulator when indexed into position. Because of this, operation proceeds on a dead-reckoning basis as the machine is effectively "blind" when manipulating a component. The image of the scene below the "eye" of the machine is digitized and stored in the supervisory computer, which then moves the turret so that the center of the field of view is over the center of the component. The television camera is geared

Figure 1. The experimental manipulator (SIRCH) used for evaluation purposes

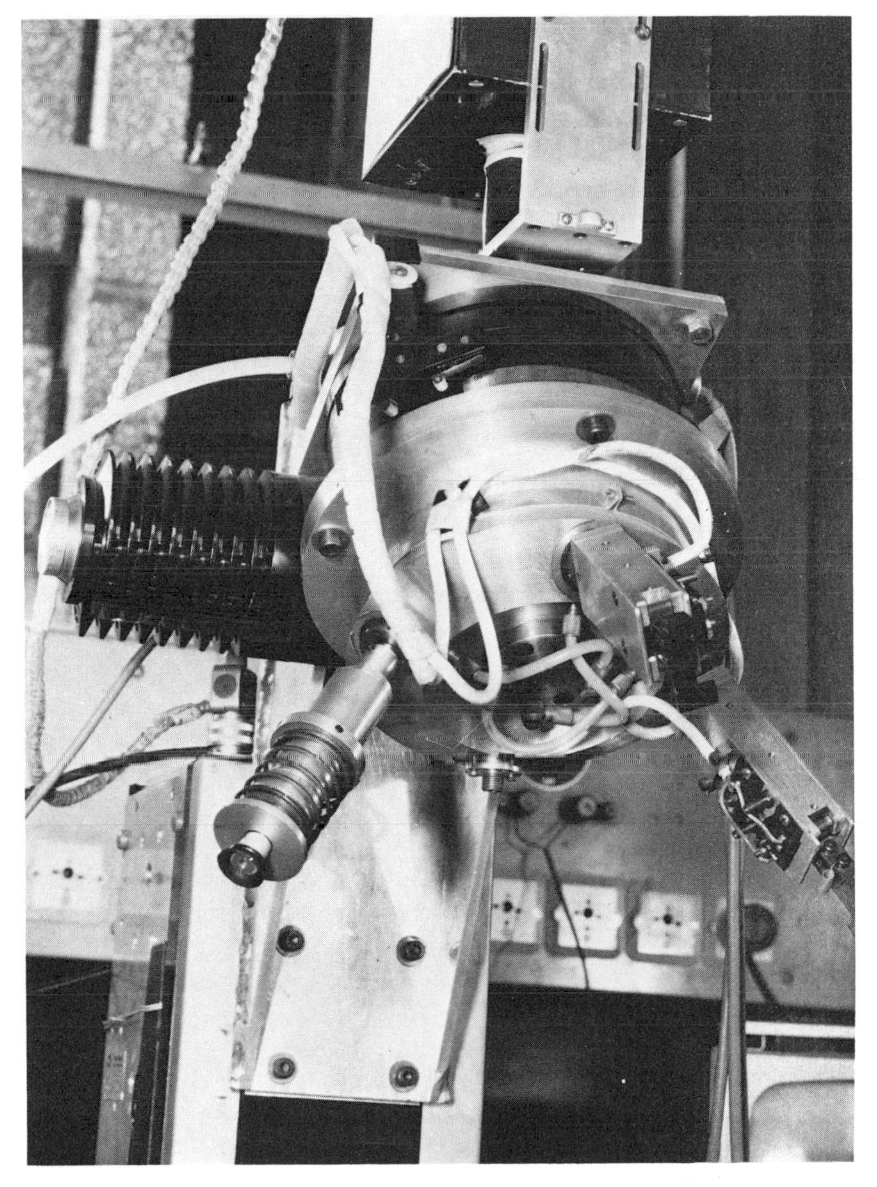

Figure 2. A close-up of the manipulating head showing the three gripping devices and the objective lens of the optical system

to the rotary axis of each of the three manipulators so that any rotation of the camera is passed on to the manipulators. This retains the angular orientation of each gripper with respect to the machine's optical axis. These features enable the machine to pick up a part by centering its optical axis over the component and then rotating the television camera to bring the image into the correct orientation. When the required manipulator is indexed into position, it is then in the correct linear and angular position for component retrieval.

The work surface on which the parts are scattered is a ground-glass plate. The scene is illuminated from below, presenting the parts in high contrast to the visual sensor vertically above. The image from the television camera is digitized into a binary scene, that is, a silhouette, for ease and rapidity of analysis. However, this means that components must be manipulated by features possessed by their peripheral edges or those of internal holes since vertical spigots and other three-dimensional features cannot be identified. In addition, parts which are heaped together cannot be distinguished from a larger object.

The grippers used for automatic manipulation are a disc-shaped vacuum sucker for picking up components that possess an unbroken flat area large enough to encompass the sucker, and a parallel-jawed gripper for grasping parts by parallel, opposite edges of the periphery or by internal holes. The choice of manipulator and the characteristics of the imaging system immediately restrict the type of component that can be handled. The component must be essentially laminar with no depressions or protrusions on its surface to foul the sucker manipulator, yet with sufficient thickness to enable the jaws of the pincer gripper to obtain a good purchase. This might be thought an unrealistic constraint initially, but in practice there are many applications that approximate to these conditions and allow reliable operation of the machine.

Manipulative Techniques

Fundamental Considerations

The field of view may contain one or more components, each of which is represented as a silhouette of its plan view. It must be remembered that a component may be several objects physically touching. The manipulators must be able to pick up one component from a closely packed array of parts without fouling themselves on adjacent objects. In addition, they must be able to grasp components reliably so that components do not fall from the grippers during handling. Finally, the action of gripping must not change the position or orientation of the grasped object. The machine is "blind" while in the process of handling parts, and subsequent operations can proceed by dead reckoning only if it can be assumed that the act of gripping does not move the component.

The manipulation algorithms operate by processing the stored representation of the "scene" presented to the assembly machine. They analyze the shape of each component with a view to applying either the vacuum-sucker gripper or the pincer gripper. The former is tried first because the analysis is simpler and therefore more rapid. If the sucker cannot be applied, then the pincer gripper is used. The algorithms try to "fit" the shape of the manipulator under consideration onto or around that of the candidate component in such a way that the gripper cannot be fouled either on other features of the part itself or by features of an adjacent part.

Preliminary Picture Processing

The supervisory computer has in its memory a digitized version of the binary image "seen" by the assembly machine. This is achieved by quantizing each of the 128 sampled lines of the television frame into 128 binary digits, thereby converting the original frame into a square matrix of 128 × 128 binary picture points. Simple algorithms are used to operate on this matrix and convert it to a form more amenable to processing but without losing any relevant information. Each component is represented by the outline of its periphery together with those of any internal holes, that is, as a line drawing. These contours then replace the original picture-point matrix in computer memory. Additional, quantitative information such as area, perimeter, centroid coordinates and maximum and minimum excursions of the contour in directions parallel to the coordinate axes is also stored for each outline. Each contour is stored in so-called "chain-code" [15] as a pair of starting coordinates and a difference sequence of direction vector. The final result of preliminary picture processing on the scene presented to the assembly machine is a number of contour sets, one set for each component, with each set consisting of the periphery plus the outlines of any internal holes. Figure 3 shows the results of preliminary processing on a simple binary scene. Note that small parts positioned inside internal holes in larger components can be accommodated without difficulty.

Application of the Vacuum Sucker

The most basic requirement for application of the vacuum-sucker manipulator is that the component should have a flat upper surface. This requirement is fulfilled as it has been stipulated that all parts can be considered as laminar. In addition, the edge of the sucker must not overlap either the periphery of the component or the edges of any internal holes when applied to the surface of the part, otherwise the vacuum cannot be maintained.

There are two fundamental advantages that accrue from the use of a manipulator of this type. The first is that the shape of the component is immaterial. Secondly, because the edges of the sucker cannot protrude beyond the periphery of the gripped part, the manipulator cannot become fouled either on an adjacent component or on other features of the part itself.

There are also two criteria that are applied to the selection of the best gripping position on a component. First, the sucker should be as far away as possible from the periphery of the object and the outlines of any internal holes. Secondly, the center of the sucker should coincide as closely as possible with the center of gravity of the part so as to minimize the turning moment due to its own weight. A sufficiently good approximation to the center of gravity is the center of area of the image of the part. These two factors act in opposite directions on the "degree of goodness" of a candidate's gripping point; the first factor is more important than the second. A figure of merit for any possible position is calculated by multiplying the minimum distance between the center of the sucker and any of the component's contours by a suitable integer weighting factor and then subtracting the distance between the sucker-center and the center of the area.

The technique that has been used to determine suitable handling coordinates consists of superimposing, in selected positions, the outline of the sucker onto that of the candidate component. If the sucker outline touches none of the part's contours at any point on its circumference, then that position of the manipulator can be used for handling. The particular set of coordinates used is selected on the basis of comparison of the calculated figure of

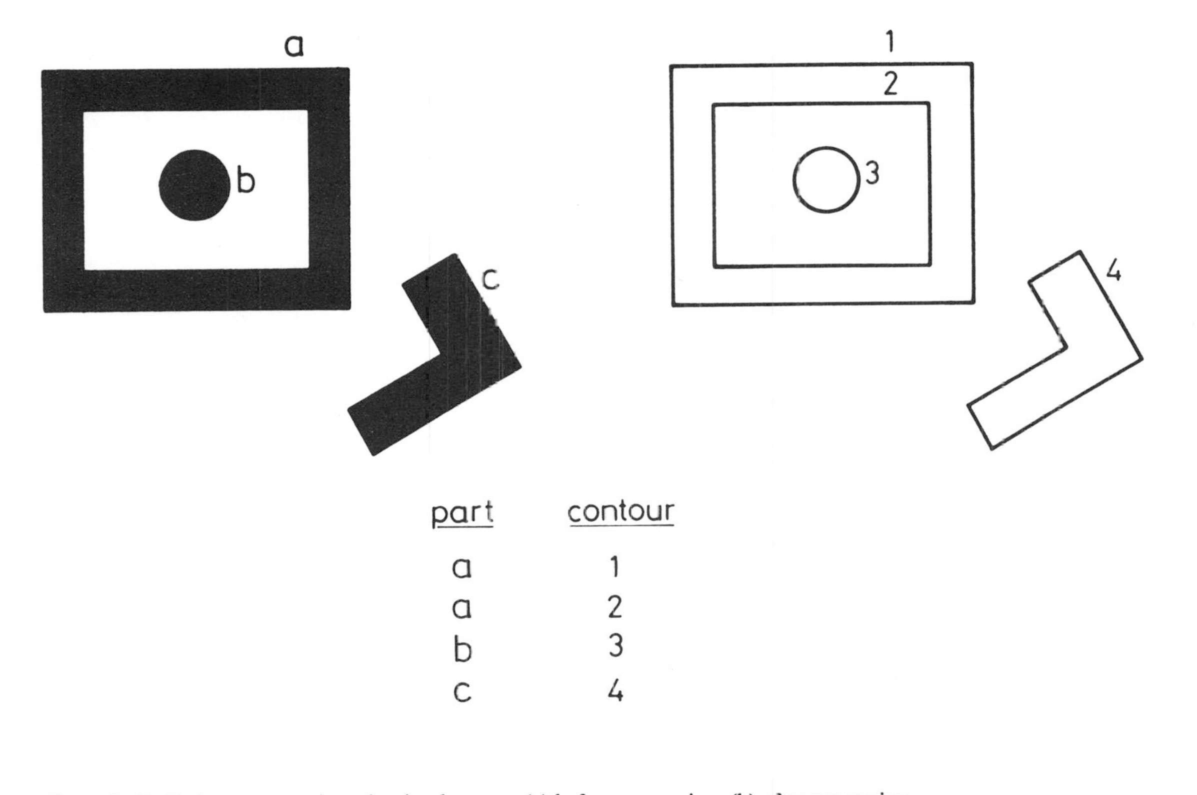

Figure 3. Preliminary processing of a simple scene: (a) before processing, (b) after processing

merit described above for each position.

Initial selection of the positions that might yield possible handling coordinates is carried out by means of a simple test that also rejects parts that are obviously too small. The enclosing rectangle defined by the maximum and minimum excursions of the component's periphery in directions parallel to the coordinate axes of the viewing system is used because this rectangle is extremely simple to compute and use. For the sucker to be considered further, the length of the shorter side of the part's enclosing rectangle must be at least as great as the diameter of the suction cup. The area within which the sucker can be tried is defined by the locus of the center of the suction cup as it traverses the interior of the enclosing rectangle with its periphery touching the sides of the rectangle. Coordinates within this area are then sampled at intervals for use as hypothetical gripping positions in subsequent tests. This is illustrated in Figure 4.

Difficulties can be encountered when examining components of particular shape and also components with large internal holes.

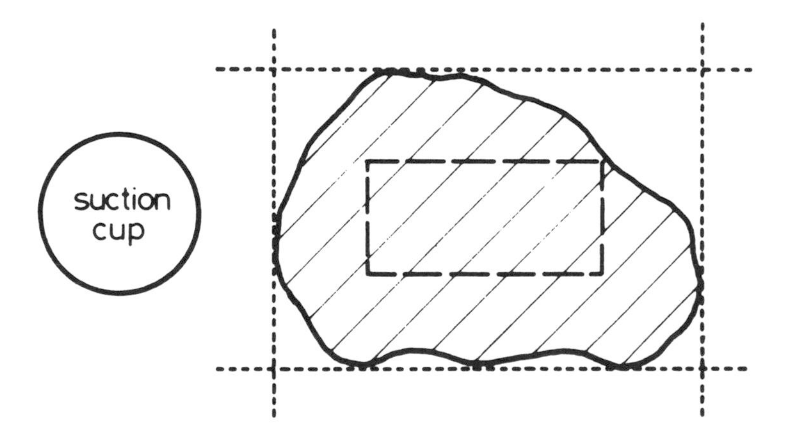

Figure 4. Definition of sampling area for suction-cup center. The sucker profile is shown on the left, with the component's enclosing rectangle and the sampling area for the suction-cup center as dotted and chained lines respectively

Examples are shown in Figure 5. In both of these cases, the positions chosen meet all the conditions required but are obviously invalid from a practical point of view. What is required is some form of test to reject all those candidate-handling coordinates not inside the body of the component. A suitable test is illustrated in

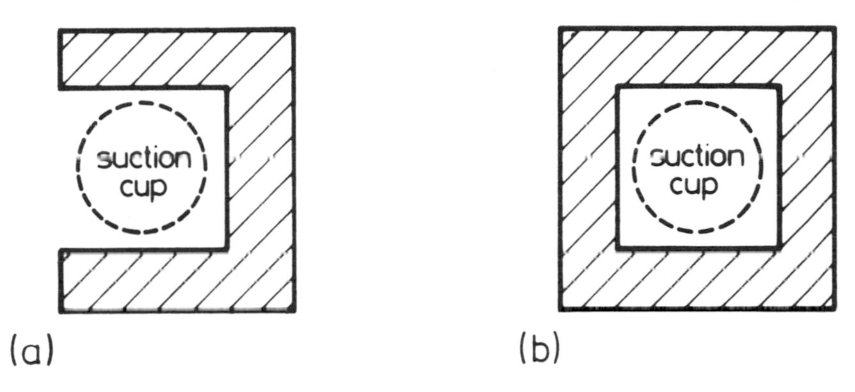

Figure 5. Difficulties caused by various component configurations: (a) re-entrant component, (b) component with a large internal hole

Figure 6. For any point, a horizontal line is drawn to the right, and the number of times that the line crosses each contour noted.

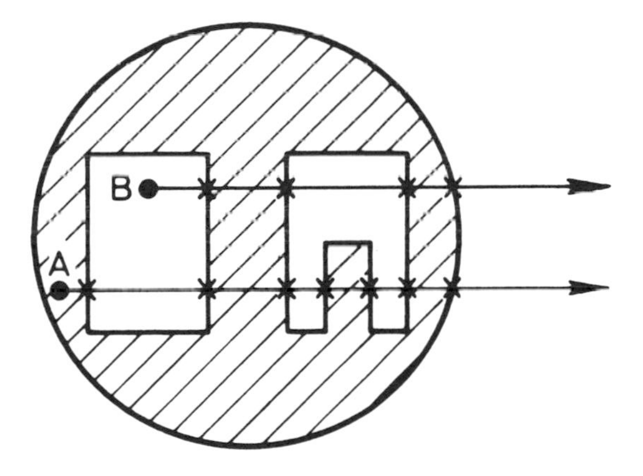

Figure 6. A test for ascertaining whether particular coordinates are within the body of a component

If the line touches the contour rather than crosses it, this fact is ignored. The number of times that the line crosses the periphery is first examined. To this quantity is added the number of contour crossings registered by the first internal hole; to this second number is added the number of crossings registered by the next hole; and so on until the final result – the total number of times that the line crosses boundaries in the line-drawing representation of the component. The partial sum is tested after examining the periphery and after each subsequent addition: it must remain odd at all times for the selected point to lie within the body of the scrutinized part. The order in which any internal holes are taken is immaterial. The only proviso is that the number of times the line crosses the periphery must be counted first.

The methods outlined above for finding suitable gripping points for the vacuum-sucker manipulator are essentially of a "cut and try" nature and consequently can be rather slow in execution if steps are not taken to speed them up. One way of doing this is to select candidate-handling points at intervals of every third picture point on every third line or row rather than at every point on every line, and also to sample every third contour point only. To this latter end, the absolute coordinates of the required contour points are calculated from the chain-coded version in a prior operation, and are then stored separately. This is included in Figure 7, which describes in flow-chart form the steps in computing the vacuum-sucker application coordinates.

Pincer Gripper Fitting

The pincer gripper consists essentially of two jaws of thin steel sheet that close to form a simple manipulator. Apart from this so-called external mode of gripping, the device can be used in an internal mode by first closing the jaws, lowering the manipulator into position adjacent to a suitable feature of the component to be picked up, and then opening the jaws to grip the component. The main features of the manipulator and examples of the two modes in which it can be used are illustrated in Figure 8. In practice the gripper is vacuum operated and the jaws are spring-loaded to ensure that approximately the same force is exerted however the gripper is used. For reliable handling, the pincer manipulator must grasp the candidate part by two opposite, parallel edges of some feature that is visible in a binary image of the part when viewed from vertically above. These edges must be straight, or approximately so, and must be long enough to provide a good purchase for the gripper jaws. In essence, the gripper-fitting process is one of template matching with the jaws of the gripper as the template. The two component edges can be selected from the contours of the candidate part in a number of different ways. Examples of these are shown in Figure 9. First, the contour of the periphery can provide

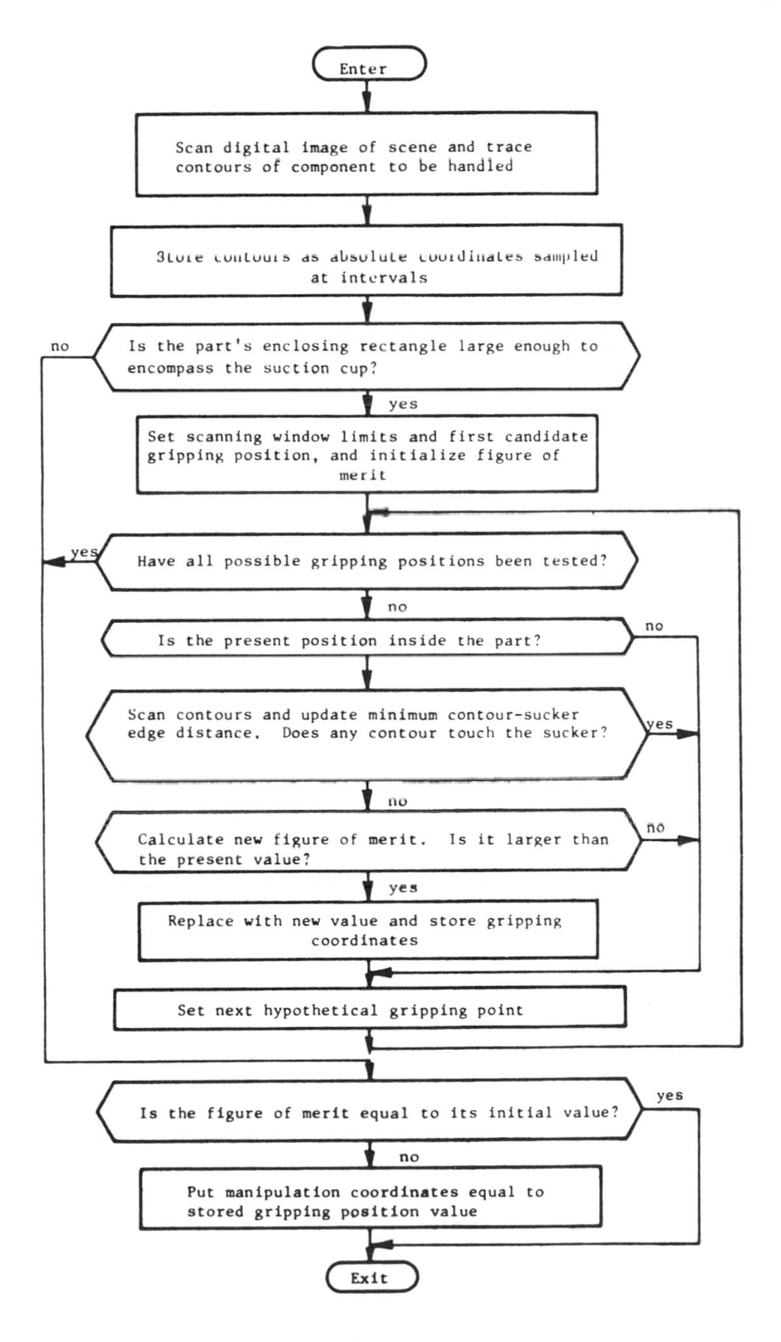

Figure 7. Flow chart for sucker fitting

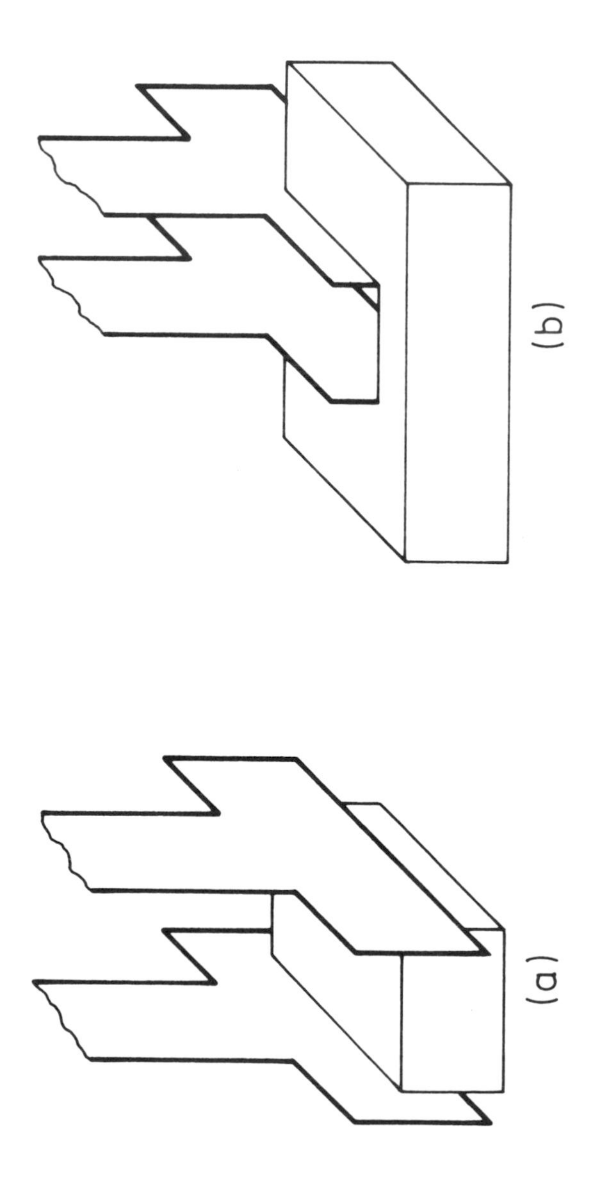

Figure 8. The pincer manipulator: (a) gripping externally, (b) gripping internally

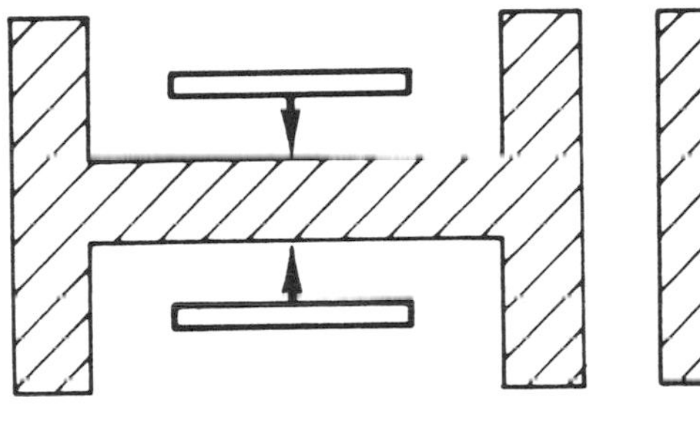

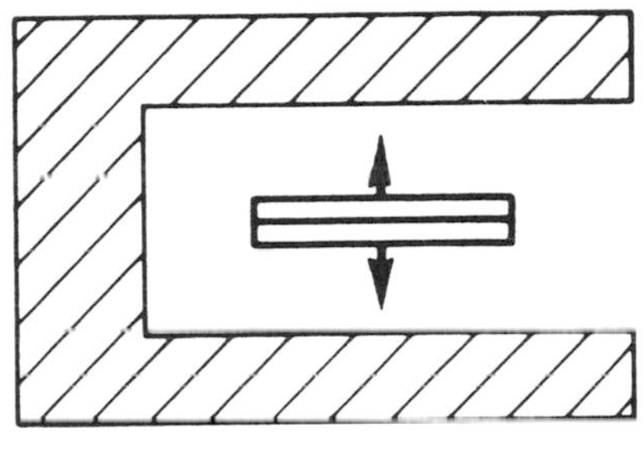

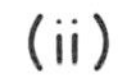

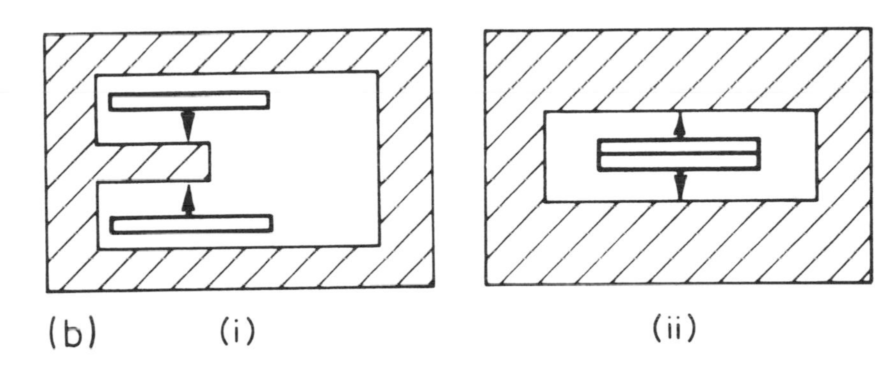

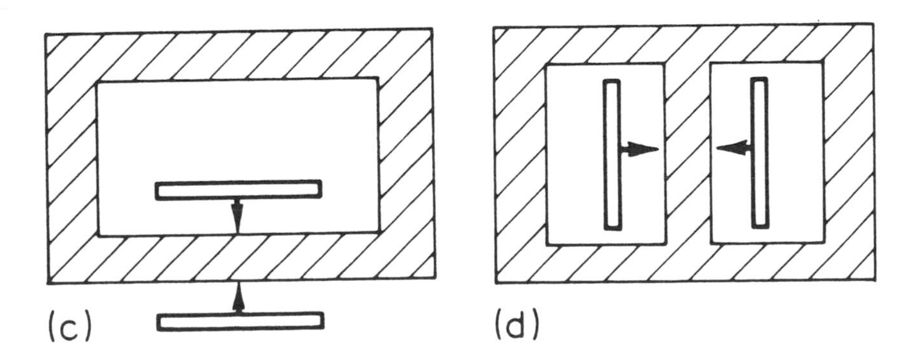

Figure 9. The possible ways of using the pincer manipulator, (a) periphery only, (i) external (ii) internal, (b) one internal hole only, (i) external (ii) internal, (c) periphery and internal hole, external mode, (d) two internal holes, external mode

both edges. Secondly, the contour of an internal hole can be used. Note that for both of these alternatives, either the external or internal mode of gripping can be used. Thirdly, one edge can be chosen from the periphery, and one from an internal hole. Finally, one edge can be selected from the outline of each of two internal holes. However, a practical difficulty arises when considering these two latter possibilities, namely that if an attempt is made to apply the gripper in the internal mode then the bottom edges of the closed jaws will strike the upper surface of the candidate part. There are ways of circumventing this problem: spring-loading the vertical axis of the manipulator, for instance; but these techniques introduce secondary problems of their own, and it is easier in the overall sense to restrict the application of the gripper in these situations to the external mode.

A number of different sets of circumstances can arise when considering a pair of component edges as possible candidates for gripping. Some examples are shown in Figure 10. When the region of overlap of the edges is longer than the manipulator jaws, no problems are encountered and the jaws are positioned centrally (Fig. 10a). When the edges overlap by an amount less than the jaw length, the overlap must be long enough to provide the gripper with a good purchase on the part. In addition, the jaws must be positioned so as to avoid other features of the component's contour. This may necessitate offsetting the manipulator so that the part is grasped by the ends rather than by the central portions of the jaws (Fig. 10b). In some cases, the manipulator jaws cannot be applied to a particular pair of edges at all (Fig. 10c). The examples shown in Figure 10 all use the periphery for grasping, but the points illustrated apply equally well to gripping by an internal hole and to periphery-hole and hole-hole gripping, except that in the latter two cases the external mode of manipulator application only can be employed. These examples also reveal that the whole area swept by the jaws from the fully open (or closed) position to the point of contact with the chosen edges must be examined to determine whether or not some portion of the contour can prevent correct operation.

The sources of obstruction to the manipulator are not limited to the candidate component itself, but can be provided by other parts in the immediate vicinity, as shown in Figure 11. Clearly, such situations must be catered for as they will occur regularly in

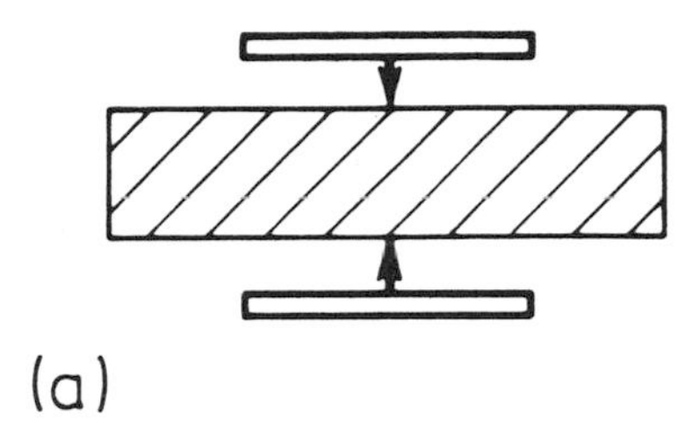

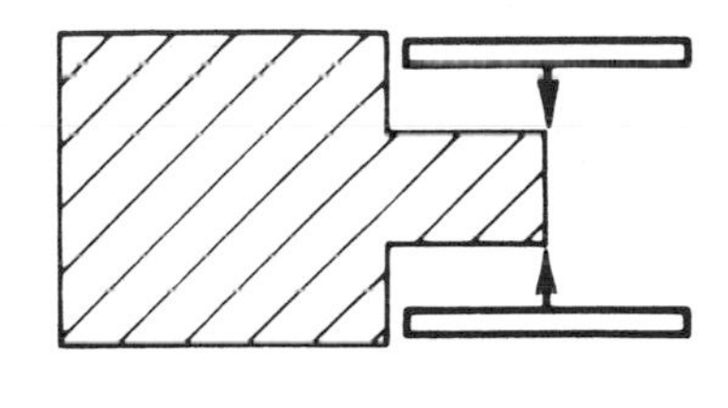

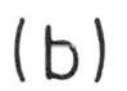

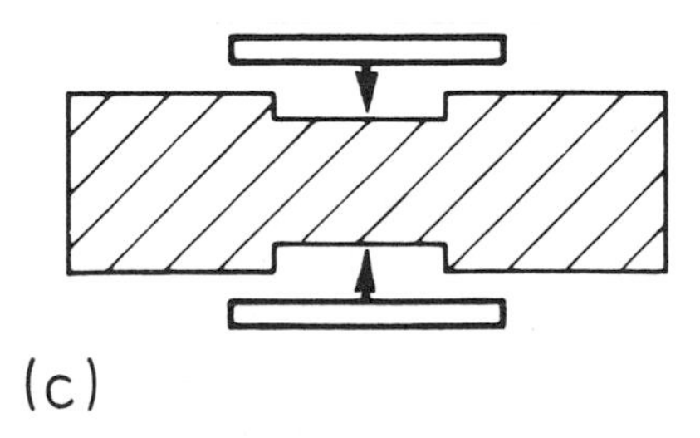

Figure 10. Situations arising when using the pincer manipulator: (a) edges longer than gripper jaws, (b) offsetting necessitated by component shape and edges shorter than jaws, (c) failure caused by component shape and edges shorter than jaws

any practical application where the components are not constrained in any way.

The first operation that must be performed in applying the pincer gripper to a part is the detection of all the straight edges present in the part's contours. As explained in a previous section, the contours are stored as **different** sequences of chain-code vectors, and it is these sequences which must be processed in order

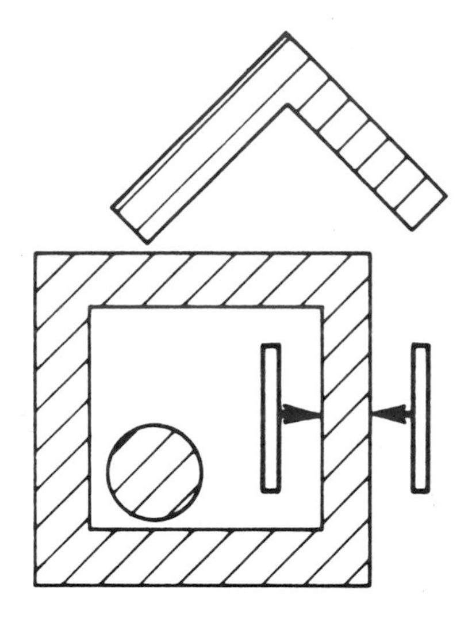

Figure 11. Obstructions to gripping caused by adjacent components. (This shows the only possible way of handling the large, square part.)

to detect straight-line segments. In practice, a piecewise-linear approximation of the contour is formed, with lines of length below a certain minimum being discarded. The overlapping of sequential lines, such as occurs when a smooth curve is approximated, is allowed since this increases the probability of each line being chosen as a possible candidate for a gripping edge. Figure 12 shows the kind of representation that results when an irregular shape is considered.

The line-extraction algorithms operate by forming two simple difference sums of the constituent vectors of a contour segment symmetrically disposed about a central vector. As long as each sum falls within a certain range, the chosen segment is approximately straight and the next pair of vectors, one at each end of the segment, are added into the difference sums. This process continues until the sums go outside the specified ranges. The start and end coordinates, the angle, the tolerance inherent in the angle of the line, and the first and last vector numbers are then stored. The

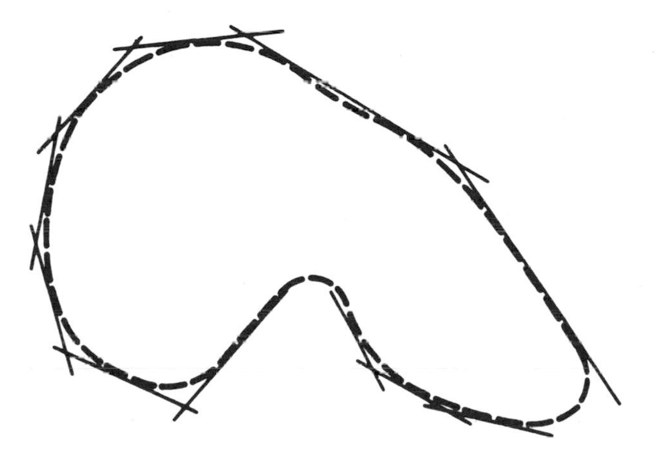

Figure 12. Line-fitting algorithm representation of an irregular shape

center point of the next line is chosen one vector point displaced from the end of the previous vector. Sequential lines whose angles differ by less than the sum of the individual angle tolerances (weighted in proportion to the length of each line) are combined into a single line. If possible, this new line is merged with the previous one and so on until two lines cannot be combined, at which point line extraction from the contour chain vectors is resumed. The line-merging process is necessary to ensure that each edge of the image of a straight-sided shape produces one line only when line-fitting is performed on it. Additional procedures are performed to eliminate redundant lines at the start and end of the contour chain (which are one and the same point).

After the conversion of the constituent contours of a part into piecewise-linear approximations, the gripper-fitting algorithms are applied. The contour or contours for consideration are chosen according to a simple hierarchical rule, namely that the periphery is considered first, then individual holes, followed by the periphery paired with each internal hole, and finally every combination of two from the internal holes of the component (in cases where there are two or more holes). The basic scene analysis and line-fitting algorithms work in such a manner that the component lines of a contour follow one another around in a counterclockwise sense. For the periphery, the interior of the closed curve formed by the straight-line approximation is that of the part itself, while for an internal hole the curve's exterior corresponds to the interior of the part. Subsequent processing can be simplified by reversing the sense of the straight-line representation of internal holes, that is, by making it clockwise. If the definition of the interior of the component is taken to be the region to the left of an imaginary observer travelling around the contour in the direction of the lines of the piecewise-linear approximation, then it can be seen that the above operation produces the desired compatibility. Figure 13 illustrates this point.

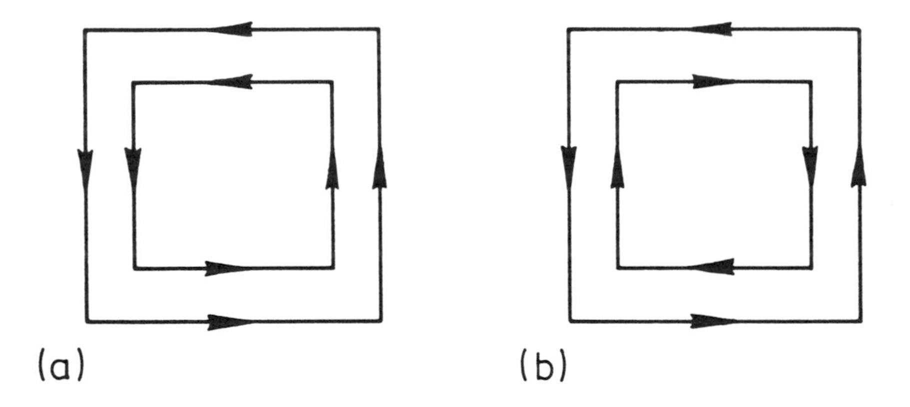

Figure 13. Reversing the sense of internal-hole contours. The arrow on each line denotes its direction: (a) before reversal, (b) after reversal

Gripper fitting consists basically of testing every possible combination of two lines selected from the contours appropriate to the scheme above, and Figure 14 shows a flow chart for the processing of the contours relevant to a single component. The most important property of suitable pairs of lines is that their angles to the horizontal should be 180° apart, that is, "antiparallel." In practice there is a tolerance associated with this condition; it is equal to the sum of the individual angle tolerances. The next most important property is that the lines overlap in the spatial sense, that is, if a normal line is drawn from each end of one line

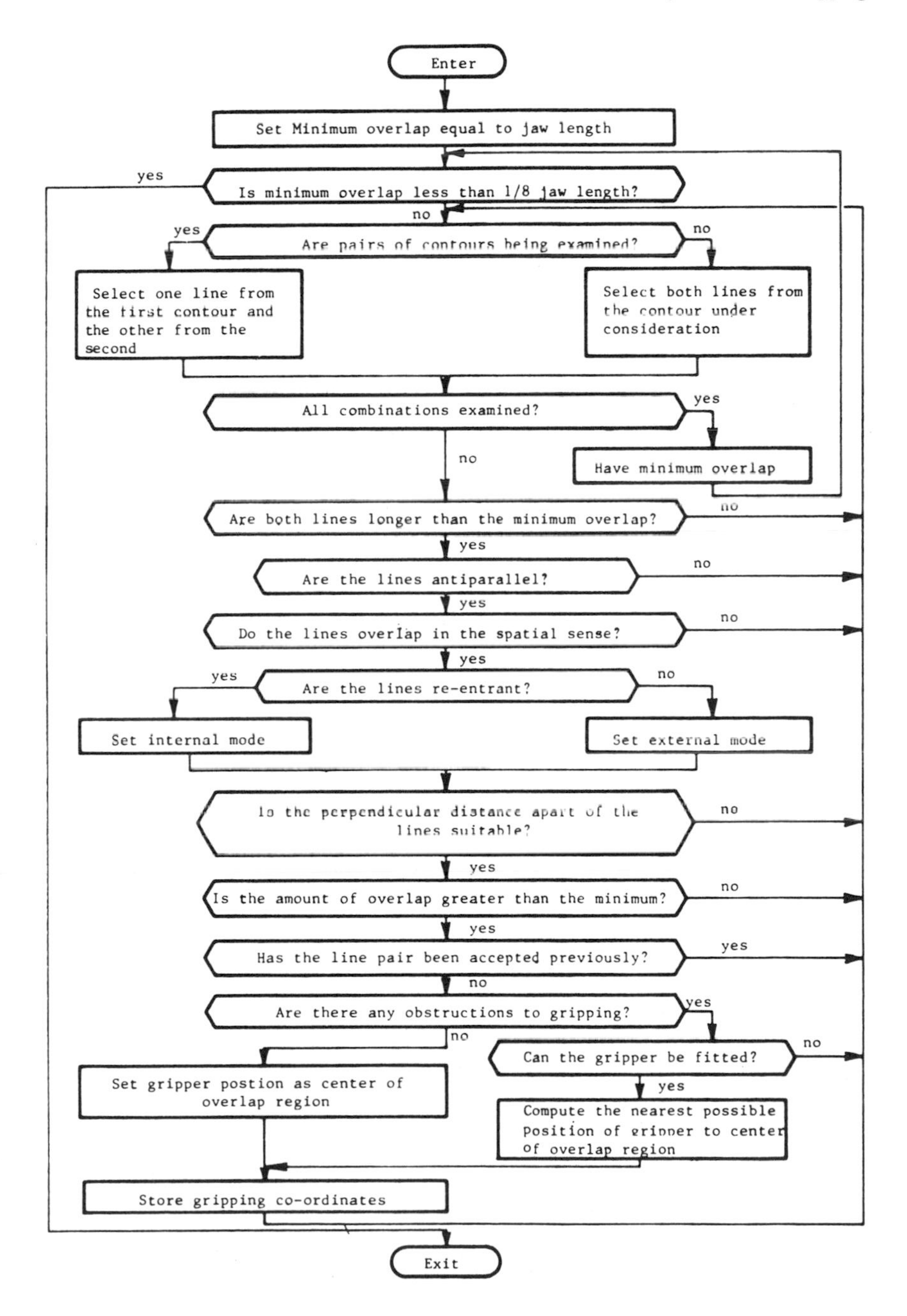

Figure 14. Operation of the gripper-fitting software

and at least one line crosses the other line, then the two lines overlap. When it has been established that the selected pair of lines may be a possible choice for gripping edges, the amount of overlap and the perpendicular distance apart of the lines must be calculated. The mode of gripping to be used, either external or internal, must be ascertained before a decision can be made on whether or not the manipulator can accommodate the chosen edges, for the mode affects the maximum and minimum permissible distance apart of the lines. This can be determined irrespective of the type of contour being considered – whether a periphery or an internal hole (these are the only two cases where the gripper can be used internally) – by noting the rotational sense of the two lines under consideration. If the sense is counterclockwise, external gripping must be used, whereas if it is clockwise the internal mode is required. This is illustrated in Figure 15.

It has been noted that the overlapping region of the two candidate lines need not be as long as the gripper jaws. Nevertheless, for practical purposes some lower limit must be set; a suitable value is one eighth the length of the jaws. For maximum speed of operation, the processing is terminated when the first position that can be used for manipulation is found, regardless of whether other, more suitable features exist. To alleviate this situation to some extent, without affecting processing time, line selection is executed in stages according to a minimum-length criterion. In the first stage, only lines longer than the jaws are considered. For the next stage or pass the minimum length is halved, and this process is repeated for each subsequent pass until the final stage (with the minimum length set to one-eighth of the jaw length) has been completed.

When a pair of lines that satisfy all the conditions for manipulation has been found, the surrounding areas must be examined for obstructions by other features of the parent component itself or by separate parts which may be in close proximity. Testing for possible obstructions involves scrutinizing the area swept by the closing or opening jaws. The algorithms construct imaginary "jaw zones" which define the swept area when the manipulator is positioned centrally over the region of overlap of the selected lines. The interior of each zone is then examined for intruding contour segments. The length of each zone is greater than that of the gripper jaws to allow for offsetting the manipulator to an unobstructed

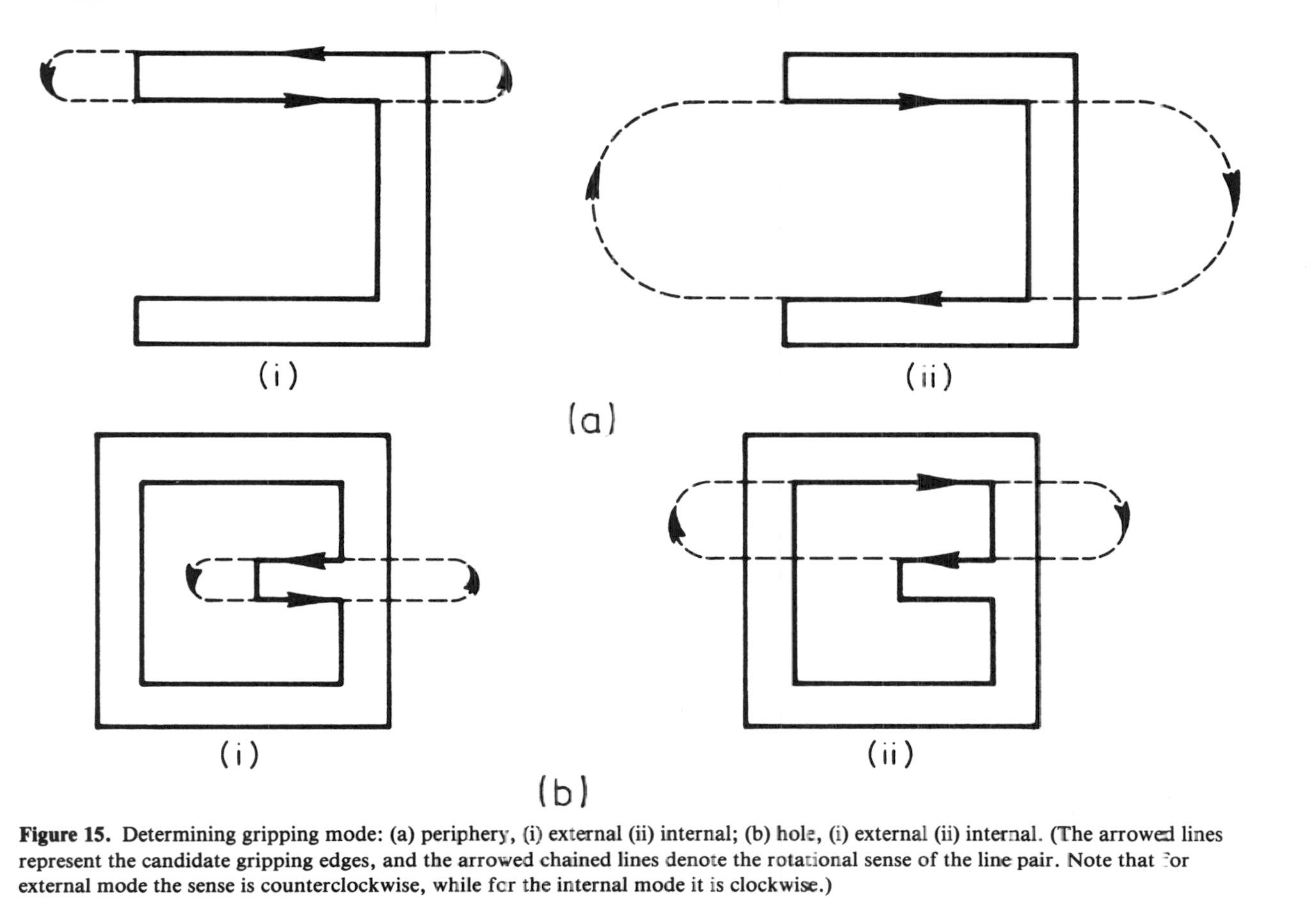

Figure 15. Determining gripping mode: (a) periphery, (i) external (ii) internal; (b) hole, (i) external (ii) internal. (The arrowed lines represent the candidate gripping edges, and the arrowed chained lines denote the rotational sense of the line pair. Note that for external mode the sense is counterclockwise, while for the internal mode it is clockwise.)

end of the overlap region. Figure 16 illustrates these points. Note that only one zone is required for the internal mode; this is because the jaws touch in the closed position (with nothing between them) and the two zones merge into one. The extension of the zones on each side of the overlap region is equal to seven eighths of the length of the jaws to allow a minimum of one eighth in contact with the region in the worst case. This figure is chosen to maintain consistency with the minimum overlap requirement.

The obstruction-seeking algorithms operate in two stages. First, the relevant contours of the component under scrutiny are examined at intervals of several chain points to ascertain if they intrude into the jaw zone or zones. In each case, those parts of the contour or contours not included in either of the chosen lines must be processed. When trying to use single-contour manipulation, both lines are derived from the same contour and two separate segments are therefore examined. On the other hand, double-contour manipulation requires the processing of that portion of each contour not included in the single straight line chosen from it. This technique is repeated for all the contours of every additional part in the field of view. The possible range of application coordinates for the manipulator, if any, is then computed. For maximum stability, the jaws are positioned as closely as possible to the center of the overlap region. Using this simple method, the manipulator can be applied (where the particular situation allows) so as to avoid fouling either nearby components or other features of the part being considered, while at the same time selecting the best position from a point of view of stability.

The Integrated System

The assembly machine described previously uses the techniques elaborated above to pick up a selection of unprogrammed engineering components scattered at random over the illuminated work area. Random disposition of the parts results in some of them being either very close to, touching, or overlapping their nearest neighbors. Adjacent components present no problems when using the sucker manipulator, and the detection of obstructions to gripping caused by other parts is an integral part of the pincer-gripper-fitting algorithms. Touching or overlapping parts look like one component, however, and an attempt to pick up this component

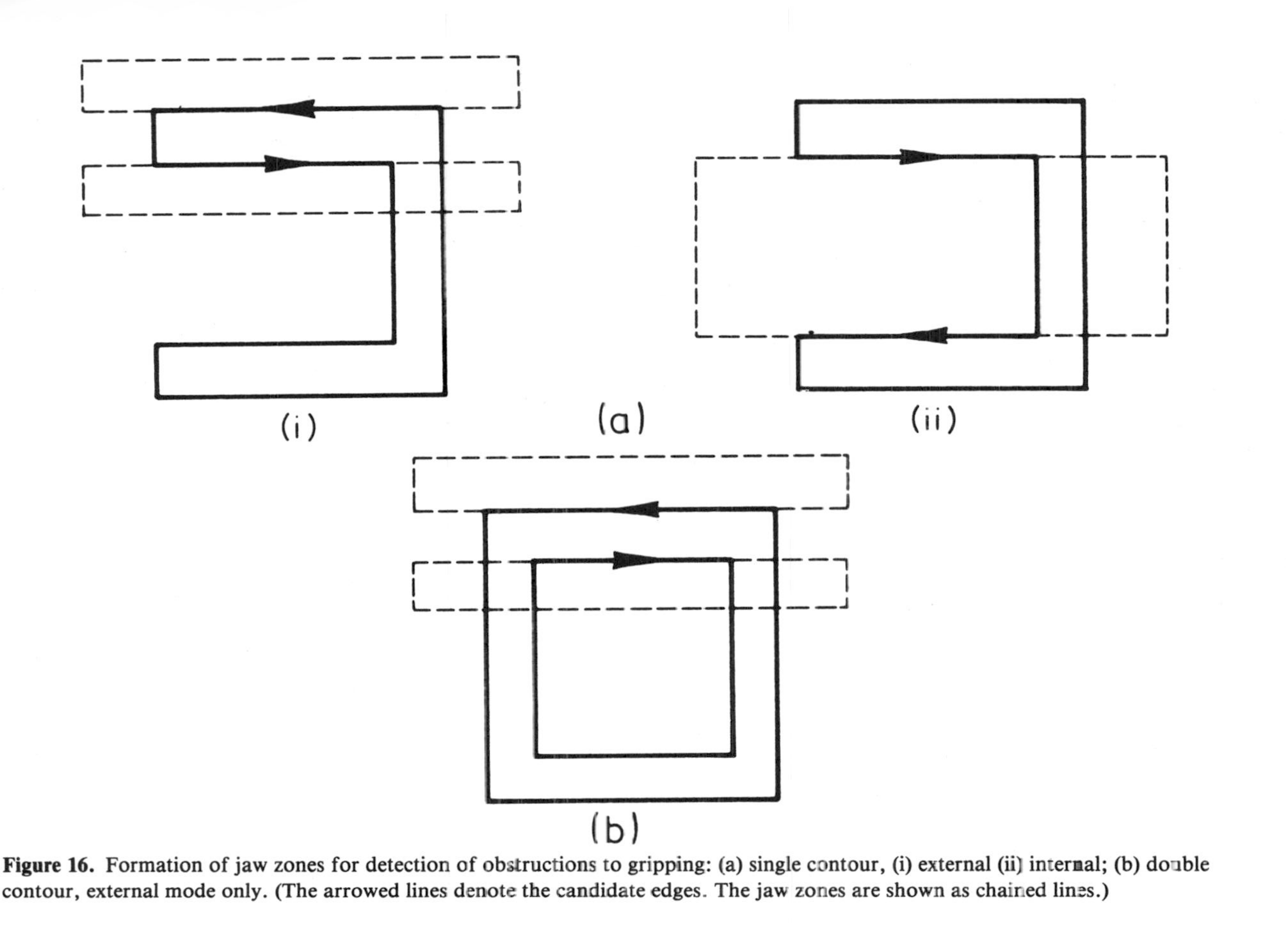

Figure 16. Formation of jaw zones for detection of obstructions to gripping: (a) single contour, (i) external (ii) internal; (b) double contour, external mode only. (The arrowed lines denote the candidate edges. The jaw zones are shown as chained lines.)

will result either in rearrangement of the whole or removal of one of the constituent parts. Subsequent attempts at handling will repeat this process until all the component parts of the compound object have been removed. This technique provides a simple and extremely effective solution to the problem of handling touching and overlapping components, and can be made an integral part of the overall procedure.

Figure 17 illustrates in flow-chart form the operation of the system. The steps involved in handling one component are shown. When operating on a heap of parts, this operation must be repeated as many times as there are components in the pile. The data which the system possesses about its application environment are limited to the physical parameters of the assembly machine's manipulators, the height of the objective lens of the optical system above the ground plane at the point where it can see the whole of the work-area (called the upper level), and the relationship at this height between the resolution of the imaging system as expressed in picture points and that of the mechanical system as expressed in steps of the linear actuators. All other information, except preprogrammed tolerances, is computed in accordance with the programmed strategy.

The upper-level scanning procedure examines each component contour by contour and stores the latter in chain-code form, together with additional quantitative information. Objects partly outside the imaging system's field of view can cause problems, however. This is illustrated in Figure 18. The scanning software rejects contours that are not closed curves; the result is that component 1 (Fig. 18a) is ignored and components 1 and 2 are seen as a single part with the internal hole of component 1 as its periphery and the periphery of component 2 as an internal hole. This problem can be resolved by placing a border of white picture points one point wide around the frame before applying the scanning procedures. The situation then becomes as shown (Fig. 18b) and the components are recognized correctly. The parameters of parts 1 and 2 are rejected after scanning because parts of their contours are imaginary.

Selection of the required component depends on the particular application of the system. The simplest method is to handle the parts in the order in which they are found during the preliminary scan, but selection on component size or basic shape, for instance,

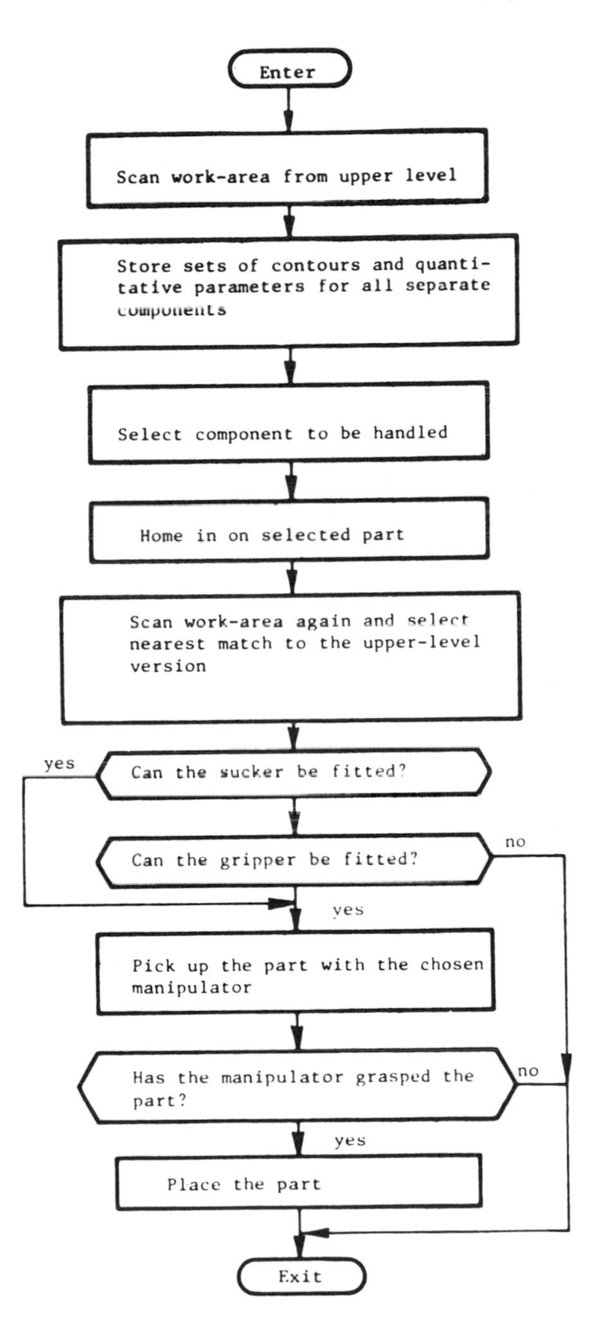

Figure 17. Flow chart for automatic manipulation

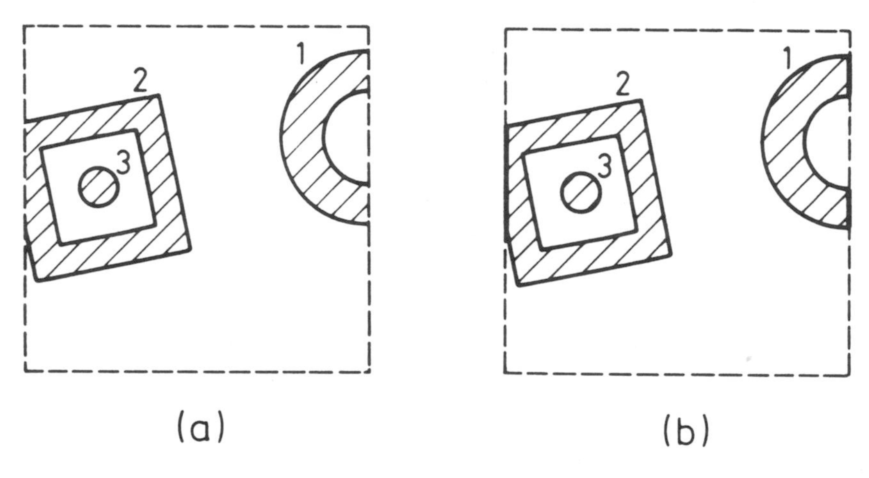

Figure 18. Problems caused by parts partly outside the field of view: (a) with no additional processing, (b) after putting a border of white picture points around the frame

may be preferable. The manipulator turret of the assembly machine is then lowered and moved in the horizontal plane so as to centralize the enlarged image of the chosen component within the field of view. The software computes the amount of vertical movement required from a combination of the scaling factor relating picture points to actuator steps at the upper level and a piece-wise linear approximation to the characteristics of the optical system. The horizontal movement required is simply converted from picture points to actuator steps.

The first task of the software, after homing-in, is to identify the image of the component to be handled from those of the parts that fall within the field of view. This situation occurs when small components are being used, the reason being that the assembly machine's manipulators would strike the work area before the required image magnification was reached. Under these conditions, therefore, several parts may be within the field of view. Identification is achieved by matching the length of the component's perimeter when seen at the lower level with the scaled version of that seen at the upper level, and by checking that the image coincides approximately with the center of the field of view. As with the upper-level processing, a border of white picture points is superimposed on the scene stored in the supervisory processor's memory, but this time the closed curves formed from partially occluded components are retained and used in the gripper-fitting tests for obstruction by adjacent parts.

After first applying the sucker fitting tests, followed by the gripper-fitting tests if the former are unsuccessful, the machine picks up the part with the appropriate manipulator applied to the computed coordinates. Correct grasping is checked by interrogating micro-switches on each of the jaws in the case of the pincer gripper or by ensuring that a vacuum is being maintained in the case of the sucker. The component is then transported to the appropriate position and put down.

The maximum size of object that can be manipulated is limited by the field of view of the imaging system and the size and lifting power of the grippers. In the present system, the CCTV camera can see an area about four inches square at the upper level. The length of the pincer-gripper jaws is one inch with a maximum gape of three quarters of an inch, while the vacuum sucker has a diameter of half an inch. A reasonable maximum size for components is therefore approximately two inches. The lower limit for the sucker is obviously half an inch, while it has been found that the pincer gripper and its controlling software can pick up hexagonal nuts of diameters down to a quarter inch with a high degree of accuracy and reliability.

Figures 19, 20, and 21 illustrate some of the more difficult situations encountered during system operation. The first diagram shows a tightly packed array of four components, numbered in order of handling by the assembly machine. The most difficult part is number 1, for which the only unobstructed gripping position is by the two edges of the lower limb as shown. Once this component has been removed, the remaining three objects are removed quickly. The approximate processing time of the whole scene when viewed from the upper level is 1.3 seconds, while that for component 1 at the lower level is 3.5 seconds.

In Figure 20, an attempt is being made to handle two overlapping discs with the vacuum sucker. The body of this manipulator is spring-loaded in an axial direction so as to allow operation on piles of over-lapping components. If part 1 is uppermost this is lifted off, leaving part 2 undisturbed, but if part 2 is uppermost then the vacuum cannot be maintained. The result is that neither component is picked up. Some disturbance of the arrangement almost always occurs, however, and it has been found that one or

Figure 19. Tightly packed array of components. (The numbers refer to the order in which the parts are handled. The gripping position for component 1 is shown in heavy line.)

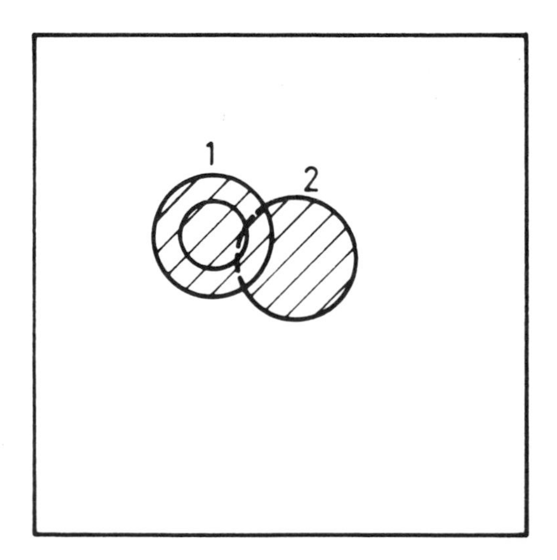

Figure 20. Overlapping components. (Part 1 is on top of part 2. The position of the suction cup is shown in heavy line.)

more subsequent attempts at handling usually prove successful.

Figure 21 shows two touching parts that are suitable for manipulation by the pincer gripper. In this case the first suitable gripping arrangement found by the software uses the manipulator internally on one edge from each of the two components. When the gripper jaws are opened, the parts are pushed apart. Nevertheless, although this attempt at manipulation proves abortive, the next two, one for each constituent of the compound "object," are successful.

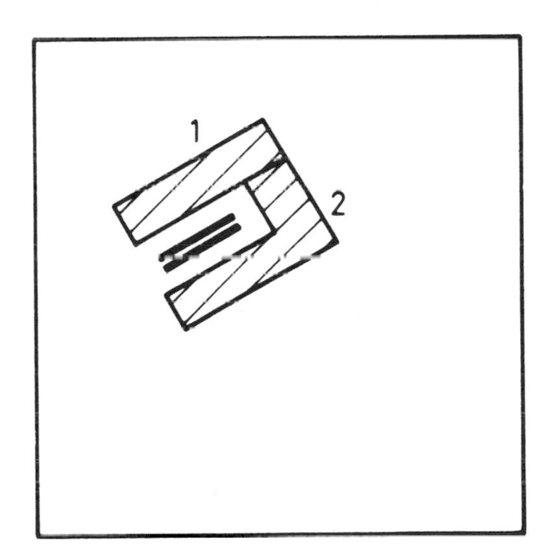

Figure 21. Touching components. (When an attempt at handling is made from the position shown, the components are pushed apart.)

Concluding Remarks

The automatic manipulation system described above has been extensively tested with many types of engineering components and has proved to be accurate and reliable. The principal reason for this success is the computational accuracy of the sucker and gripper-fitting software. Even though the resolution of the imaging system is not high (128 X 128 picture elements), it has been found possible to handle reliably some comparatively small components and also many with intricate shapes. A particularly important property of the software is its speed of operation. Other existing systems exhibiting a similar degree of sophistication are sometimes slow in comparison with a human operative. This system, however, compares favorably in "processing time" with an untrained person in some situations; for instance, in determining how to pick up a part with a relatively complex shape. The system is also compact in terms of processor memory: the software, including data areas, can be accommodated in 6 kilowords of the 16-bit memory of the host computer.

Two particularly significant points have arisen as a direct result of this research. The first is that a robotic system such as that described here, which makes decisions about the problems at hand according to a programmed overall strategy and which computes all the parameters it needs from the application environment itself, is far more reliable in operation than a system which operates by matching preprogrammed information to the application itself. The second point is that although the techniques presented in this paper are simple and are based largely on common-sense considerations, when observed objectively they imbue the host machine with a disproportionate amount of intelligence. While it is reasonable to conjecture that this can happen only in very restricted situations, it indicates nevertheless that similar methods for increasing the versatility of other industrial robots are feasible.

The industrial application implicit in the previous discussion is automatic assembly using sensory feedback, in which automatic manipulation is essentially an enhancement to the system's capabilities. There are some applications, however, in which an automatic handling facility alone can suffice. Examples are simple sorting and palletizing tasks, where parts must be lifted off conveyer belts and arranged on trays or loaded into magazines. It might be thought that the limiting of components to those with an essentially laminar format is too restrictive for practical purposes, but experience has shown that there are many instances where a two-dimensional view does not compromise accuracy or reliability.

Acknowledgments

The mechanical configuration of the handling system described in this paper was developed jointly with Professor W. B. Heginbotham of the Production Engineering Research Association, Melton Mowbray, England. His collaboration with this work is greatly appreciated.

Some parts of the research program have been supported by the Science Research Council, London, England.

References

1. Winston, P.H., The M.I.T. Robot, in *Machine Intelligence* 7, Meltzer, B. and Michie, D. (ed.), pp. 431-63, (Edinburgh University Press, 1972).
2. Duda, R.O. and Hart, P.E., Experiments in scene analysis, Proceedings of the First National Symposium on Industrial Robots, I.I.T. Research Institute, Chicago, April 1970, pp. 119-130.
3. Forsen, G.E., Processing visual data with an automaton eye, in Pictorial Pattern Recognition, Proceedings of Symposium on Automatic Photo-interpretation, pp. 471-502, (Thompson, Washington, D.C., 1968).
4. Barrow, H.G. and Crawford, G.F., The Mark 1-5 Edinburgh Robot Facility, in *Machine Intelligence* 7, Meltzer, B. and Michie, D. (ed.), pp. 465-480, (Edinburgh University Press, 1972).
5. Pugh, A., Heginbotham, W.B. and Kitchin, P.W., Visual feedback applied to programmable assembly machines, Second International Symposium on Industrial Robots, I.I.T. Research Institute, Chicago, May 1972, pp. 77-88.
6. Heginbotham, W.B., Page, C.J., and Pugh, A., A Robot research at the University of Nottingham, Fourth International Symposium on Industrial Robots, Japan Industrial Research Association, November 1974, pp. 53-64.
7. Yoda, H., Ikeda, S., and Ejiri, M., A new attempt at selecting objects using a hand-eye system, *Hitachi Review,* 22, Part 9, pp. 362-5, 1972.
8. Tsuboi, Y., and Inoui, T., Robot assembly using tv camera, Sixth International Symposium on Industrial Robots, University of Nottingham, March 1976, pp. (B3) 21-32.
9. Gleason, G.J., and Agin, G.J., A modular vision system for sensor-controlled manipulation and inspection. Ninth International Symposium on Industrial Robots, sponsored by the Society of Manufacturing Engineers and the Robot Institute of America, Washington, D.C., March 1979, pp. 57-70.
10. Ward, M.R., Rossol, L., and Holland, S.W., Consight: A practical vision-based robot guidance system. Ibid., pp. 195-211.
11. Kelly, R., Birk, J., Duncan, D., Martins, H., Tella, R., A robot system which feeds workpieces directly from bins into machines. Ibid., pp. 339-355.
12. Dessimoz, J.D., Hunt, M., Zurcher, J.M., and Granlund, G.H., Recognition and handling of overlapping industrial parts. Ibid., pp. 357-366.
13. Page, C.J., Visual and tactile feedback for the automatic manipulation of engineering parts, Ph.D. thesis, University of Nottingham, U.K., 1974.

14. Heginbotham, W.B., Page, C.J., and Pugh, A., A practical visually interactive robot handling system, *The Industrial Robot,* Vol. 2, No. 2, 1975. pp. 61-66.
15. Freeman, H., Techniques for the digital-computer analysis of chain-encoded arbitrary plane curves, *Proc. Nat. Electronics, Conf.,* Vol. 17, 1961.

Chapter 8

An Interface Circuit for a Linear Photodiode Array Camera

D. J. TODD

Natural Environment Research Council
Computing Service, England

Abstract An interface circuit is described for connecting a linear photodiode array camera to a microcomputer. The circuit uses shift registers to overcome the data rate difference between the camera and the computer, and can easily be extended in both the number of bits of digitization and the number of pixels in a line.

This paper describes an interface circuit to connect a linear array photodiode camera to a microcomputer. The camera is being used as part of a triangulation range finder for a mobile robot. It uses a 64-element photodiode array, made by Integrated Photomatrix Ltd (IPL), mounted on a board carrying the array driver circuit and a simple output amplifier. The board is shown in Fig. 1; the photodiode array is the integrated circuit near the center. This board, which is supplied complete by IPL, gives a signal with an amplitude of a few volts. The dynamic range is about 20:1 with this particular output circuit, so the brightness can be adequately digitized as a 4- or 5-bit word. In the circuit described here 4 bits are used, but the word length could easily be increased by adding extra shift registers.

The interface circuit works by reading a line (i.e., 64 picture elements) from the photodiode array, digitized to 4 bits by an analog to digital converter, into a 64 long by 4 wide shift register, then shifting the line out of the register and into the computer at a lower rate. This procedure is necessary because the minimum data rate of the array (about 50,000 values/sec) is too fast for the microcomputer to cope with directly. The circuit (See Fig. 2.) can be used with any computer having at least one input and one output port. The machine used here is a Cromemco Z2D; a Z80 based, S100 bus microcomputer.

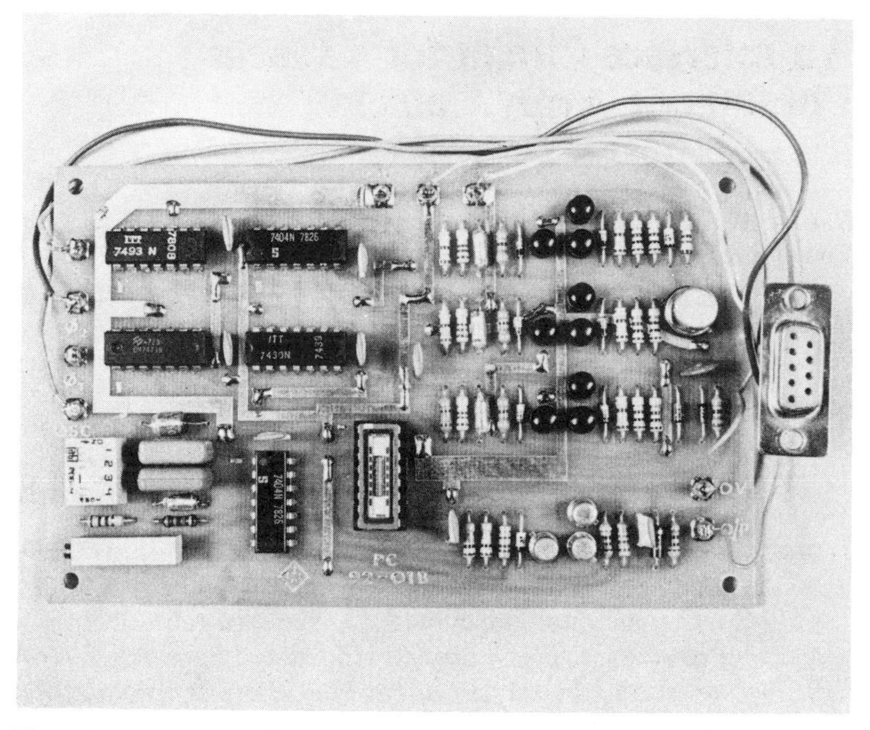

Figure 1. Printed circuit board of camera. The photodiode array is the integrated circuit near the center

Circuit Operation

The computer commands the circuit to store a line from the camera by sending a low pulse on one bit, the "start" bit, of the output port. This clears flip-flop 1 in readiness for the next SCAN pulse from the camera. When this SCAN pulse is received from the camera flip-flop 2 is set, allowing clock pulses to be received by the clock input of each MC14517B shift register. The 4-bit value of a pixel is transferred to the shift register on the rising edge of the signal $\overline{\text{OSC.Q2}}$.

About 1 μs after this edge, the "start conversion" signal for the next pixel to the ADC goes high. Conversion then takes place in the next 8 cycles of the 7413 clock. The ADC is a MicroNetworks MN5213, which can convert to 8 bits in 6 μs. Shifting of data into the shift register continues for the 64 pixels, until ended by the next SCAN pulse. Once flip-flop 1 has been set (i.e., Q1 is low) no more SCAN pulses will be accepted until the computer does another

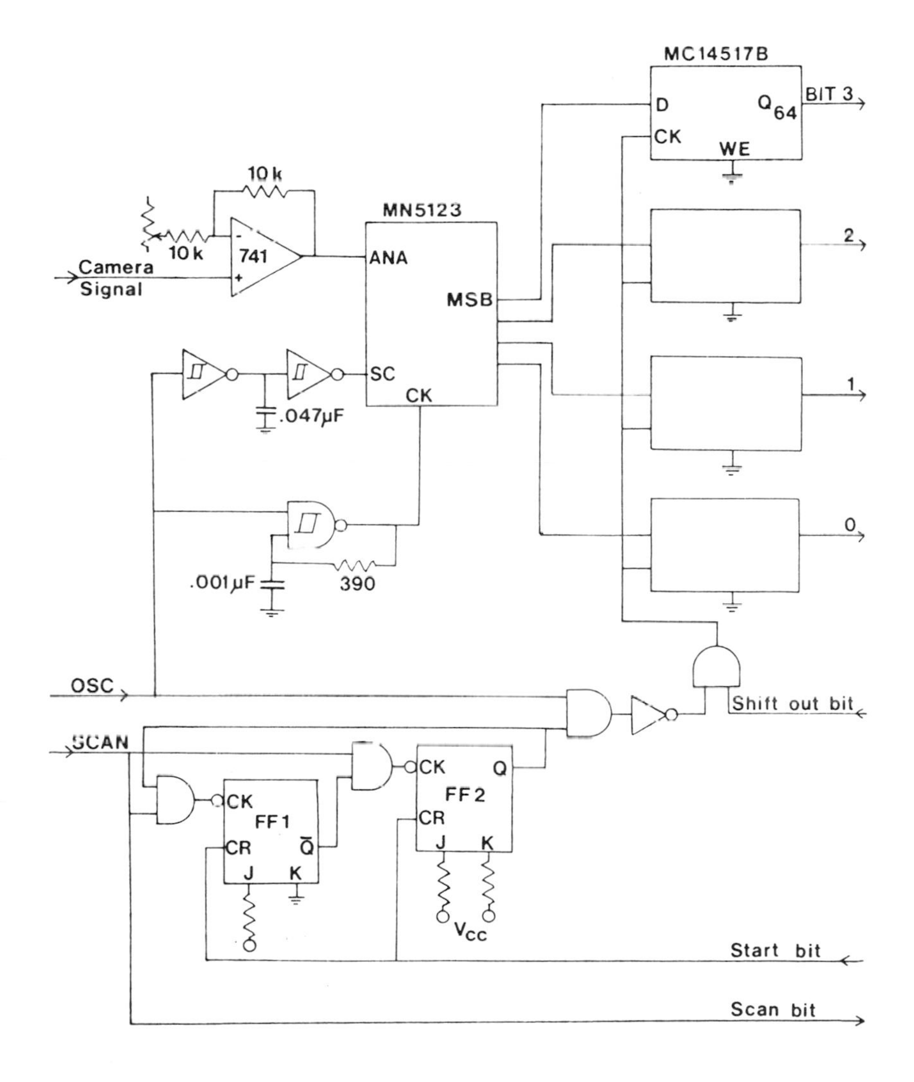

Figure 2. Interface circuit. Both flip-flops are 7473; all AND gates are 7408; the Schmitt trigger circuits are 7414 and 7413. The shift registers are Motorola MC14517B. The signals OSC and SCAN are generated by the camera

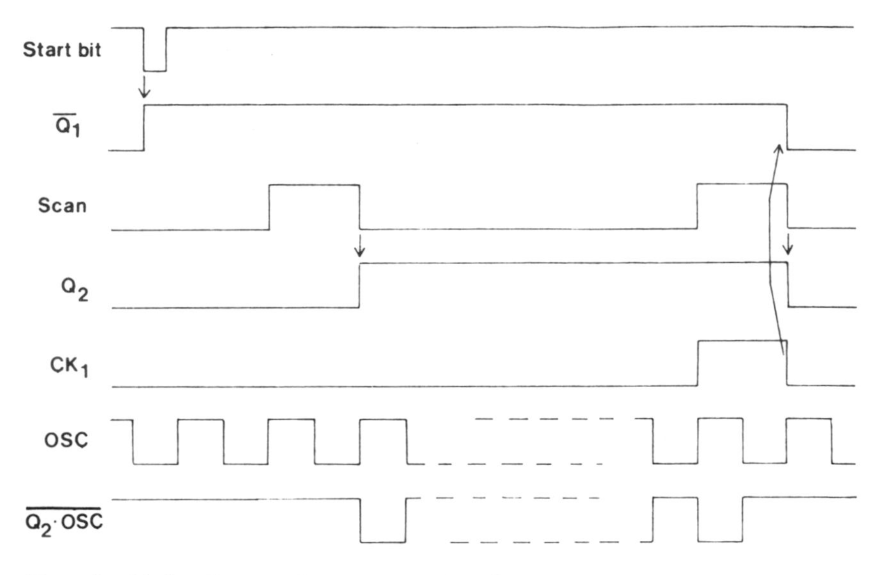

Figure 3a. Timing diagram of camera-to-register phase

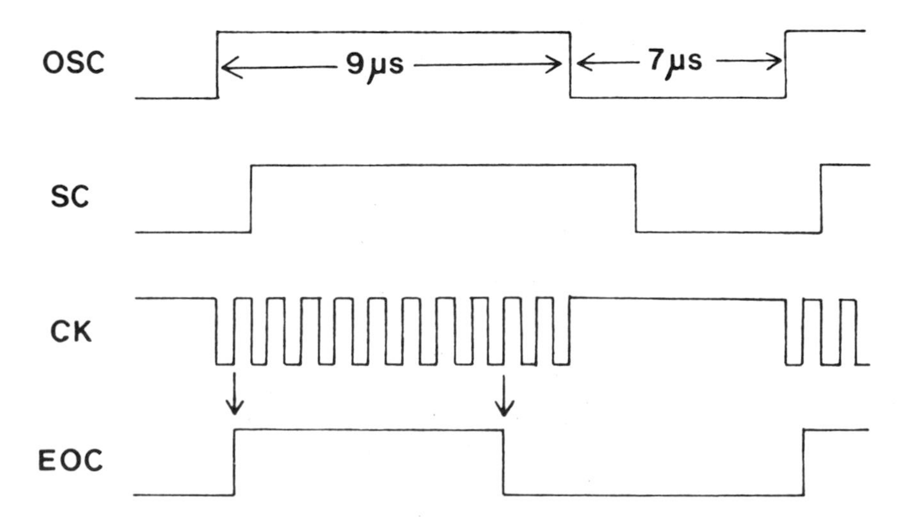

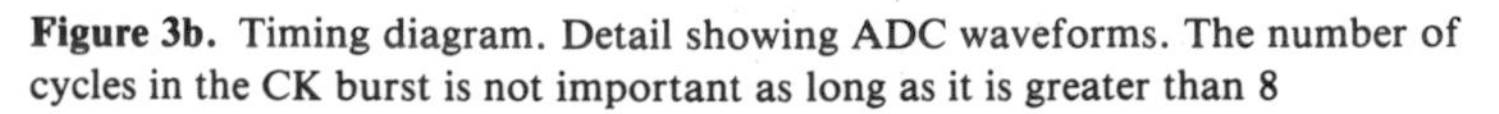
Figure 3b. Timing diagram. Detail showing ADC waveforms. The number of cycles in the CK burst is not important as long as it is greater than 8

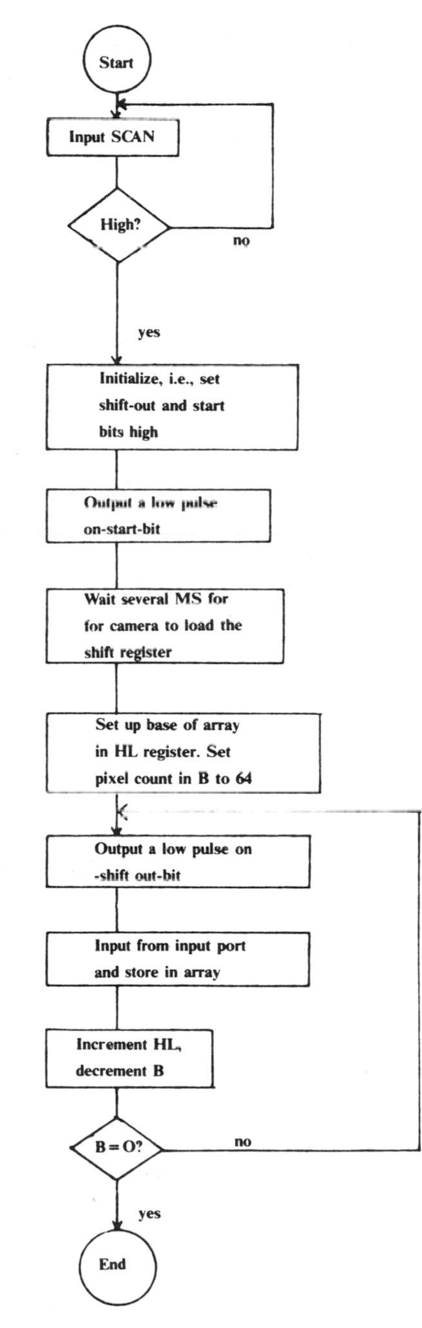

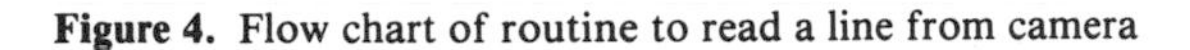
Figure 4. Flow chart of routine to read a line from camera

"start" bit output. Therefore, the line of data will remain in the register until the computer chooses to read it out. To do this it outputs a low pulse on another bit, the "shift out" bit, of the output port, repeating this cycle 64 times.

Programming

The flow chart of a subroutine to read a line from the camera is shown in Fig. 4. In order to synchronize with the camera scanning cycle, the computer must wait until the SCAN signal goes high. On detecting this, the computer gives the "start" pulse, then waits long enough for the camera to shift a complete line into the shift register. It then does its 64 input cycles, storing each 4-bit value in one element of an array.

An example of a 64-element line image obtained using this interface is shown in Fig. 5. The central peak is produced by light reflected from a narrow white object.

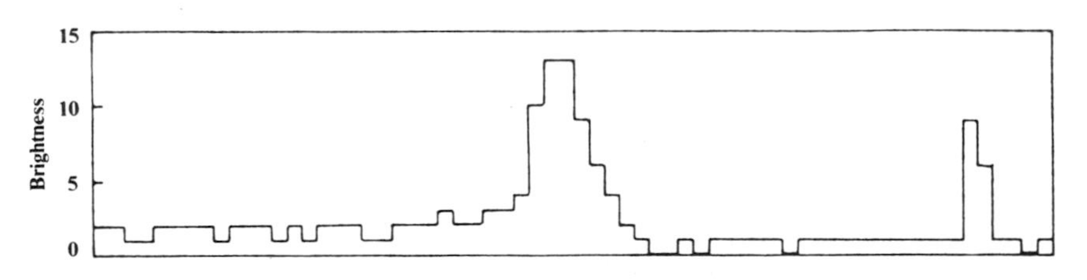

Figure 5. Example of line image. The vertical axis represents brightness in arbitrary units. The peak in the center is the image of a narrow white object. The smaller peak is a spurious signal at the end of the line

Components Used

The photodiode array is the IPL 4064; the board it is mounted on is the K4064. The lens is an F1,9, 25 mm television lens.

Acknowledgments

This work was supported in part by the Whitworth Foundation. I should like to thank Dr. A. H. Bond for the use of the facilities of Queen Mary College Artificial Intelligence Laboratory.

Part III:
Adaptive Processing for Vision

Chapter 9

Networks of Memory Elements: A Processor for Industrial Automation

T. J. STONHAM

Department of Electrical Engineering & Electronics
Brunel University, England

Abstract The classification of non-deterministic data, with particular reference to visual information, is examined. The problem is established by reference to pattern recognition of alphanumeric characters. Networks of memory elements are presented as adaptive learning processors, and their application to the recognition of images subject to varying degrees of constraint is discussed. Pattern detection in a data base, where the relationship between the data and their classification is unspecified, is demonstrated with reference to medical pattern recognition. The implications of pattern detection in general process control are considered.

Introduction

Automation is, in essence, the implementation of decisions by a machine, on data which are obtained from the environment on which the machine is to operate. The data may take many forms, and be derived from various sources. Visual images, discrete measurements, and processed measurements in the form of spectra are typical examples.

At the outset, automation appears to be an exercise in instrumentation; this is a valid description where a machine—be it a mechanical device, a digital computer, or some other electronic processor—is required to perform a monitoring and controlling role on its environment, and where the control strategy can be specified.

A pump and float switch can be employed to control automatically the level of water in a tank. A typical arrangement is shown in Fig. 1. The essential feature of the system is that it is deterministic. The data—the liquid being at, or lower than, a predefined level—can be adequately measured, and provide suitable information for the machine (the pump/switch system) to operate on under normal circumstances.

If, however, under extraordinary conditions, the rate of inflow to the tank is greater than the pump capacity, the system will fail. Other system malfunctions can arise through component failures, blockages, etc., which can themselves be monitored provided they are anticipated at the design stage. Thus the accuracy of the system, in terms of maintaining a constant level, is a function of system complexity.

While any deterministic task can be automated at the expense of complexity, unacceptably large operating times might be incurred on a conventional computer structure where large amounts of data are involved. In practice, however, most operations leading to some form of decision on data extracted from a real environment cannot be fully specified, and are at best fuzzy and at worst implemented by human intuition.

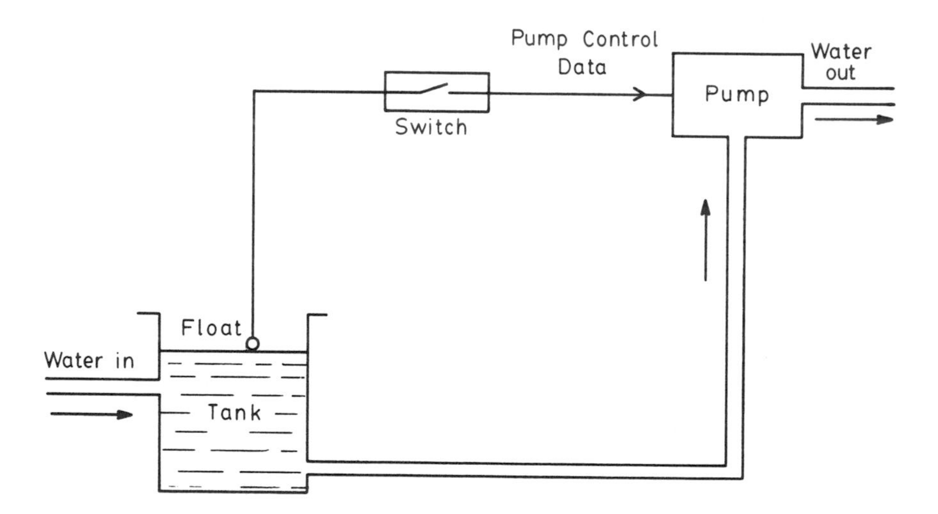

Figure 1. Automatic level control

The performance of the human is generally regarded as a reference point in assessing the success of an automated task. However, the undoubted ability of the human in the field of pattern recognition, together with a lack of understanding of the mechanics whereby a human is able to process visual data, has led to an underestimation of the problems inherent in pattern recognition.

Character Recognition—A Deterministic Problem to a Greater or Lesser Extent

Alphanumeric character recognition was one of the earliest subjects for machine implementation. The data are readily classifiable by a human. The reader can identify characters as he reads this paper, although he is likely to be processing the printed information at a more sophisticated level than if he merely assigned labels to individual characters. The success of alphanumeric character recognition depends largely on the nature of the data. Single-font readers have been operating successfully for many years, and manufacturers claim an accuracy approaching 100%. The problem is, however, highly deterministic, as the characters have been suitably stylized and precisely printed. The shape of individual characters can, therefore, be uniquely defined, and detected by a series of tests, usually in the form of a mask-matching procedure. While automation as defined at the beginning of this paper would appear to have been achieved in this application, the cost/complexity penalty has been shifted to the printing procedure where high tolerances have to be observed. A reduction in overall costs could be achieved by permitting poorer quality printing. The data would consequently become less precise, and require a more powerful pattern-recognition procedure to maintain the performance level.

Recognition of unconstrained machine printed characteristics, as found on merchandise labels, addresses on envelopes, and the printed media in general, represent data in which a formal description of each data category becomes increasingly difficult. It is possible that all the fonts could be described, although regular updating would be necessary as styles constantly change. Furthermore, the quality of print from an impact typewriter spray and electrostatic printers is generally poor, and considerable deviation from a specimen of a particular font can occur. If a spatial comparison between character and prototype is made, ambiguities can arise due to different characters being closer in terms of a point by point comparison than different versions of the same character, as can be seen in Fig. 2.

Recognition of machine printed characters is in essence feasible. The human has no difficulty in identifying the data in widely differing forms and subjected to extensive corruption and distortion, thereby suggesting that each data category does contain some common distinctive information, although it is not necessarily definable. The problem is, however, becoming less deterministic, and the fuzziness of the data is considerably greater than with single-font OCR type characters. The application of specific rules and algorithms becomes more complex and time-consuming. Thus the need for a more generalized approach becomes apparent. Ideally, one needs to be able to measure the degree of "n-ness'' present in a character to be classified, where n takes, in turn, the value of all the data categories in the classifier and the ''n-ness'' factor is independent of style, font size and orientation. As this factor cannot be precisely defined, implementation of a classifier on a general-purpose

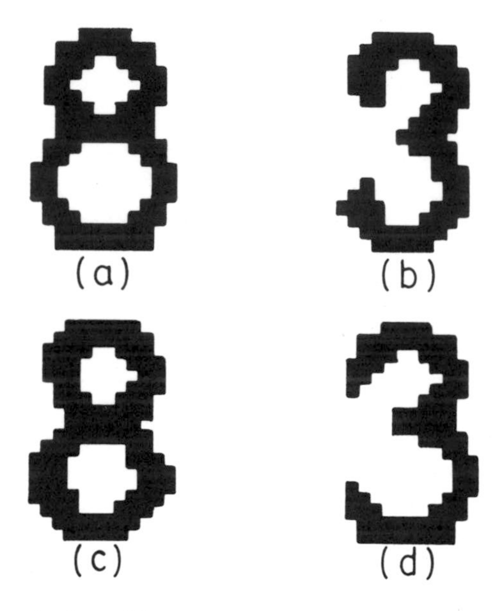

Figure 2. An example where different numerals are more similar than versions of the same numeral. A point-by-point comparison reveals (a) and (b) to be closer than (a) and (c) or (b) and (d)

computer using conventional programming techniques becomes less tractable, and has led to the development of adaptive learning techniques whereby a function is evolved by exposing the adaptive classifier to a known representative set of the data to be encountered. The resulting function will, ideally, be sensitive to the overall features of its particular data category. Networks of memory elements will, in the subsequent sections of this paper, be shown to provide a facility for learning and generalization.

The recognition performance for unconstrained machine printed characters approaches 98% correct per character [1]. However, if the data are in the form of a code or label of, say, 10 characters with no redundant information present, the expected accuracy per code will fall to 80%, a less impressive figure, illustrating the very high performance figures which in certain applications must be achieved.

Hand-printed characters represent a non-deterministic form of alphanumeric data. No two examples of the same character are ever identical, and a given character may satisfy more than one category descriptor, as shown in Fig. 3. The need for a generalized approach becomes imperative with this form of data, and commercially viable machine recognition has only been achieved when the recognition is carried out at the time of writing, thereby enabling the operator to correct any errors. The author's experiences with these devices revealed a need to conform to a range of preferred styles in order to achieve a barely acceptable performance. In effect, the diversity of the input data was being reduced.

Cursive handwriting is the most difficult form of alphanumerics to

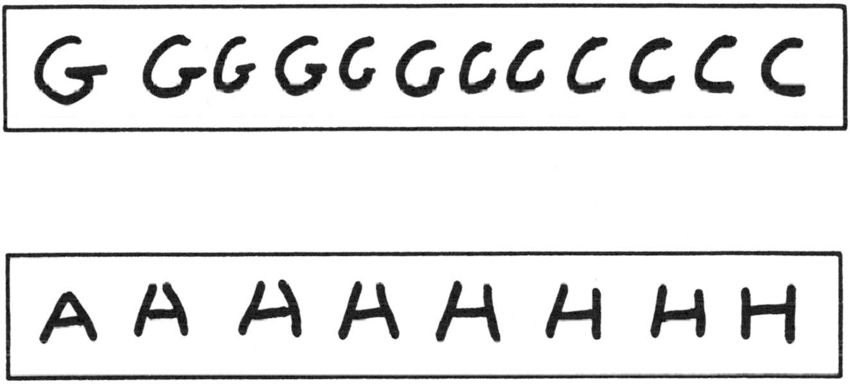

Figure 3. Ambiguities in hand printed characters

read by machine. Individual characters, when isolated, are, in some cases, unrecognizable by the human, who relies to a greater extent on context than with other alphanumeric forms. Any automation of this form of data capture will of necessity require constraints to be applied before any recognition is attempted. The need for automation in this area is nevertheless a real one, an example being the transcribing of archival material from documents to electronic storage media, where the volume of data precludes a manual approach.

Some Aspects of Industrial Pattern Recognition

In considering data processing tasks which are more industrially oriented, a pattern of performance and achievement, dependent on the nature of the data, follows lines similar to those outlined in the character recognition overview. The numerically controlled machine tool operates in a precisely defined universe, as does the dedicated robot. In these examples, simple in terms of data processing, a series of operations are carried out as a function of time. The operations are precisely defined by algorithm or by example (the latter in the case of paint-spraying robots which duplicate a manual initialization procedure). The system is deterministic, implements a single set of tasks, and cannot adapt to any changes in its environment unless they have been foreseen at the design stage and incorporated into the system.

In automation involving digital processing of visual images a requirement for very high resolution may have to be met. An alphanumeric character can be adequately represented at a two-dimensional resolution of 16 × 16 pixels, whereas mechanical parts machined to high tolerances may require a resolution of the order of 1000 × 1000 pixels. Any operation which requires the examination of the majority of pixels in an image at this resolution will be very time-consuming if a Von Neumann type of computer structure is employed. The operating time is likely to be unacceptable, despite the fact that the rate at which physical objects are presented to a camera must be slow (at most 10 per second) in order to maintain physical stability. Furthermore, microprocessor devices are not suitable for high-speed, high-capacity data manipulations. Therefore, in certain applications, parallel processor architecture must be considered, even if the operations on the data which give rise to a decision are in themselves quite simple.

The need to process visual images occurs in inspection and quality control exercises and in automatic sorting and assembly procedures. Image processing is also required in robots with visual feedback systems.

The number of data categories can range from a simple dichotomy, for example, an acceptable/unacceptable decision in an automatic quality controller or the verification of a signature at point of sale, to perhaps many tens of thousands in a sorting or stock-control application.

Operations on visual images of machined metal parts may be regarded as highly deterministic. However, the effects of spatial quantization and electrical noise from the camera and data acquisition system will give rise to distortions. A greater degree of uncertainty arises from the image of an object being orientation dependent. In two dimensions (sheet metal parts on a flat surface) an unconstrained part can appear in an infinite number of orientations, and the problem is compounded with three dimensional parts which can also have an unknown number of positions of gravitational stability. The problem on closer consideration appears to be stochastic, and the algorithmic approach to decision making becomes less appropriate.

There exists a further class of problems where human experience plays a large part in the formulation of a decision, but the actual rules are not defined (and in some cases perhaps not definable). An operator who is familiar with a machine tool can detect when, for example, a cutter is blunt, by the noise and vibrations made by the machine. Experienced persons can detect abnormalities in electrocardiograms, and electroencephalograms can recognize passing vehicles from seismic signals. A wide range of decisions are made on data bases where the precise rules and operations leading to the decision have not or cannot be defined. In such cases a model exhibiting some aspects of human learning and adaptation is likely to be more effective than an exhaustive algorithmic assessment of the data in order to establish adequate data/classification relationships.

The following sections of this paper will consider how pattern recognition, the essential decision-formulating procedures for automation, can be implemented using networks of memory elements and applied to deterministic problems, fuzzy data, and situations where a learning procedure provides the only practical means of establishing a relationship between data and their appropriate classification.

Networks of Memory Elements for Pattern Recognition

Given a low resolution pattern matrix of 16 $\times$ 16 binary pixels, the number of different patterns which can be represented within that space is to all practical purposes infinite (of the order of 10^{76} patterns). A pattern recognizer connected to the input matrix is required to detect a subset of the 10^{76} patterns. The definition of the subset may not, however, be fully specifiable, as can be illustrated with an alphanumeric example. If the character A could be specified by a prototype, with an allowable distortion tolerance of, say, 5%, the character A could be readily detected in an input space using a bit by bit comparison with a prototype. If less than 5% of the corresponding pixels between unknown pattern and prototype differed, the unknown pattern could be classed as an A. This rule would be able to identify 10^{22} patterns as the letter A, but it would not recognize all patterns that look like an A. Further inadequacies of this Hamming distance approach are shown in Fig. 4.

In handling and sorting piece-parts which are precisely machined,

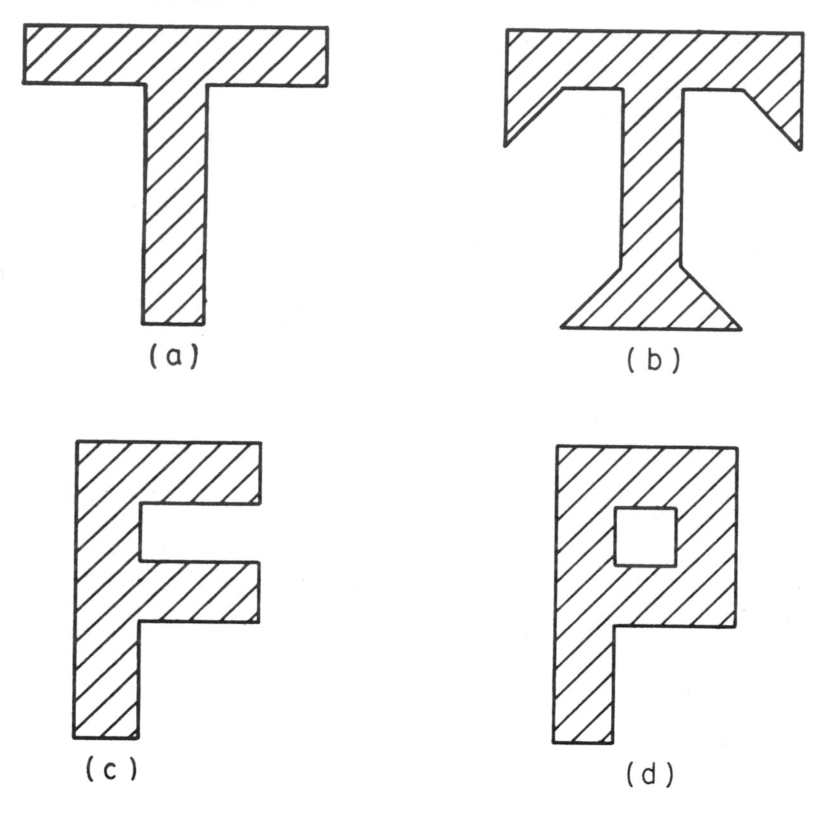

Figure 4. Examples of characters which would be misclassified by a "5% derivation from prototype" rule
If (a) is prototype T, pattern (b) would not be classified as a T
If (c) is prototype F, pattern (d) would be classified as an F

cut, or cast, prototype patterns can be accurately defined and distortions would be expected to be within specified tolerances. However, freedom of orientation on a conveyor belt or work surface, together with distortions arising from any electro-optical system producing the digital patterns on which the recognition is carried out, will degrade the performance of prototype matching or algorithms operating on discrete measurements. The "familiar object viewed at an unusual angle" problem (Fig. 5) can often baffle the human, and illustrates the wide range of aspects which may be encountered in viewing a precisely defined part.

In other applications, for example, the monitoring of the quality of a seam weld, the problem would prove to be intractable on a prototype description basis.

A learning system obviates predefined pattern prototypes or descriptors. The learning system to be discussed here has its origins in the Bledsoe and Browning [2] n-tuple method, first proposed some twenty years ago. It has, however, only recently become possible to implement the technique economically in either hardware or software at reasonably high data resolution.

Figure 5. Two different images of a familiar object

Let the universal set of patterns that could occur on a binary matrix be U. Several sets A, B, C, etc. within that universal set need to be identified. Let a representative sample of, but not all of, the patterns within sets A, B, C, etc., be available to the designer. These patterns form training sets for the learning system. Ideally, one requires a generalization set G_i which can be interpreted from the training set. However, depending on the nature of the training set, the system may undergeneralize, and not recognize all the patterns in sets A, B, C; or it

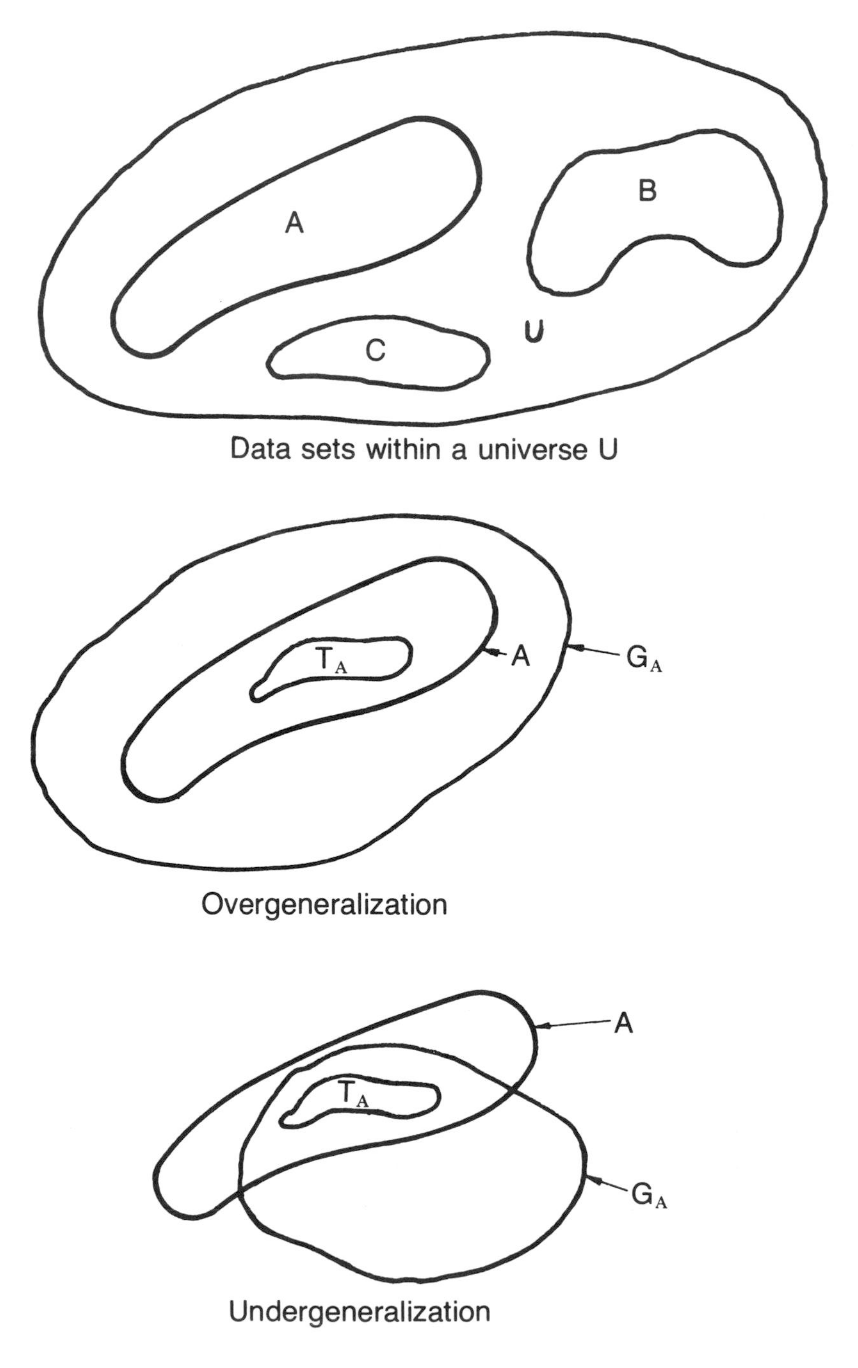

Figure 6. Set representation of patterns

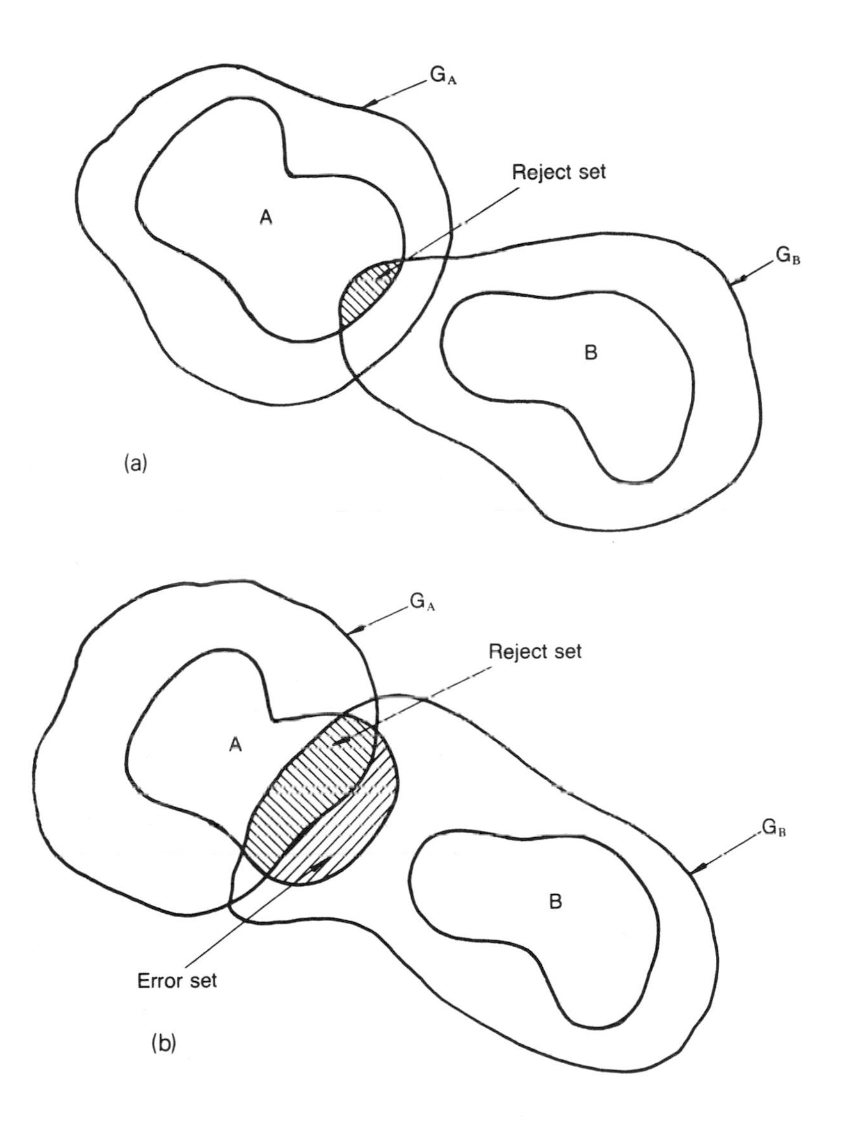

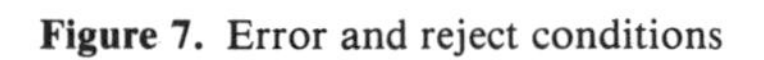
Figure 7. Error and reject conditions

may overgeneralize and accept spurious patterns outside the data sets (see Fig. 6). It is important to note that the relationships between patterns within a set pattern and between sets are not necessarily linear functions of Hamming distance [3] [4].

It is unlikely that precise correspondence between G_A and A will be achieved, and overgeneralization with respect to a given data set can be tolerated (and is indeed desirable) provided that generalization sets of different data categories do not overlap, causing patterns to be identified with more than one training set (a reject condition; see Fig. 7a), or a generalization set of one category overlaps another data set which is not fully covered, through undergeneralization due to an inadequate training set (an error condition; see Fig. 7b). A fuller discussion of generalization properties is given in [5].

A generalizing processor can be implemented with a single layer network of memory elements, as shown in Fig. 8. Each memory element, having a n-bit address field, samples an input matrix and extracts either at random or in some predetermined way an n-tuple of pattern displayed there (in this case n = 3). The n-tuple subpattern provides an address for the memory and a flag (logical 1 if the stores have been initially set to logical 0) is stored, to indicate the occurrence of the particular value of subpattern being sampled on the input matrix. A single layer network of memory elements, which will be referred to as a discriminator, is exposed to a representative set of patterns from a given class of data, and a discriminator function is set up in the memory elements by flagging the appropriate locations of memory addressed by

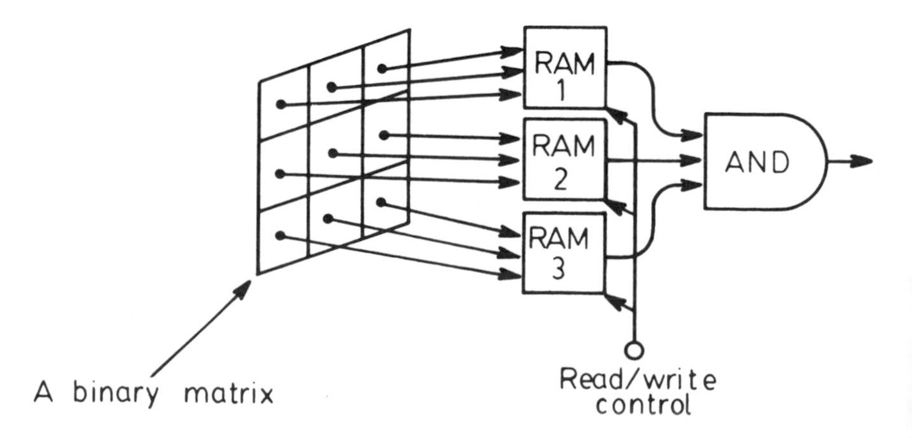

Figure 8. A discriminator comprising three memory elements

the subpatterns obtained from the input space. The discriminant function is derived solely from the data, and requires no *a priori* class description or rules. Furthermore, there is no direct storage of pattern data, and the amount of storage in a discriminator is independent of the size of the training data set.

Having trained a discriminator, an unknown pattern to be classified accesses the stored data in the memory elements—again using subpattern values as address data for the memories, and a decision is made on the outputs. In the simple arrangement in Fig. 8 an AND function is used. However, other functions including a numerical summation of the outputs, can be employed.

The generalization properties of the discriminator in Fig. 8 can be illustrated in the following example:

Given training set T_A as

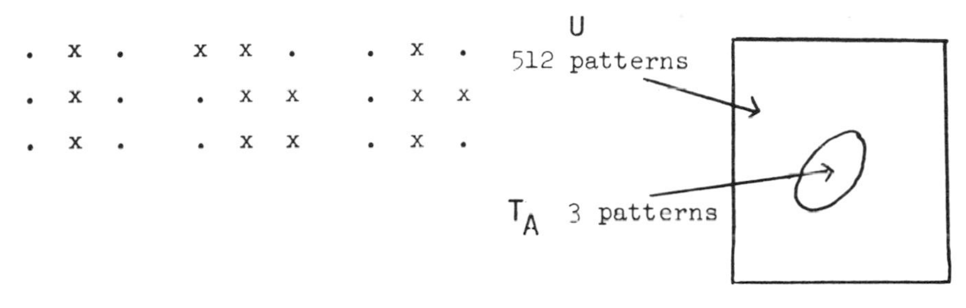

The following patterns are in G_A in addition to T_A:

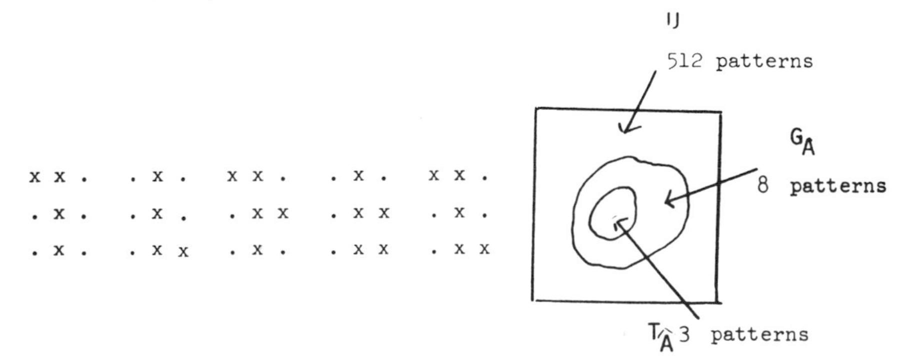

The size of the generalization set is given by:

$$|G_A| = \prod_{j=1}^{N/n} k_j \tag{9-1}$$

where k is the number of different subpatterns seen by the jth memory element during training, n is the n-tuple subpattern size and N is the input size. Only patterns in G_A will stimulate a response from the discriminator in Fig. 8 trained on T_A.

A classifier would contain a number of discriminators. In a simple configuration there would be a discriminator for each data category. Implementation can be either as a software system or in hardware using semiconductor memories (RAMs or ROMs)*. The latter can be organized serially or in parallel. The WISARD pattern recognition device under development at Brunel University has parallel discriminators and serially addressed memory elements, and will operate at its highest resolution 512 × 512 pixels at a rate of 4 images per second. A hardware system schematic is shown in Fig. 9.

The storage requirements S per discriminator increase exponentially with n-tuple size and linearly with input resolution:

$$S = \frac{M.X.Y}{n} \quad 2^n \quad \text{bits} \tag{9-2}$$

where X is the horizontal resolution, Y is the vertical resolution, M is the mapping ratio between input and discriminator (usually 1) and n is the n-tuple subpattern size. Storage requirements are summarized in Fig. 10. (It should be noted that 10^6 bits of storage can be accommodated on a single printed circuit board, at a chip cost of about £350.)

Applications

Networks of memory elements configured as learning pattern recognizers have been applied to a wide range of problems including character recognition, spectral identification, medical diagnostics, and automatic fault detection in digital equipment ([5] and references cited therein). In this section emphasis will be placed on the recognition of visual images of piece-parts using single layer networks of memory elements.

The separability of different data categories does depend on Hamming distance, although this dependence is not linear in n-tuple recognizers. A rule of thumb is (not surprisingly): the more alike two different objects are, the more difficult it is to distinguish between them. Exam-

*RAM: Random Access Memory; ROM: Read-Only-Memory.

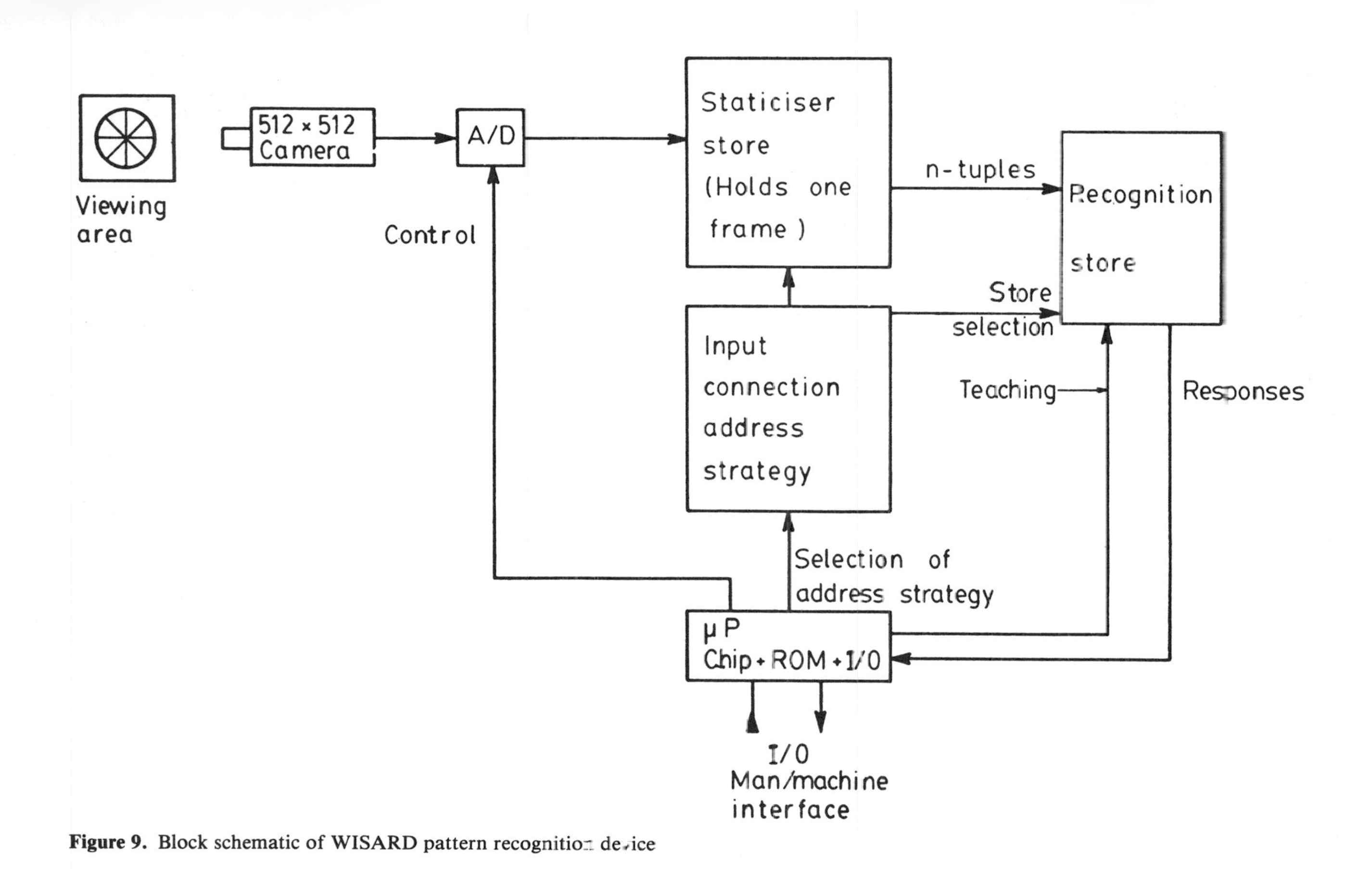

Figure 9. Block schematic of WISARD pattern recognition device

ples will therefore center on similar parts requiring different identifications, and categories exhibiting gross differences are assumed to be readily distinguishable.

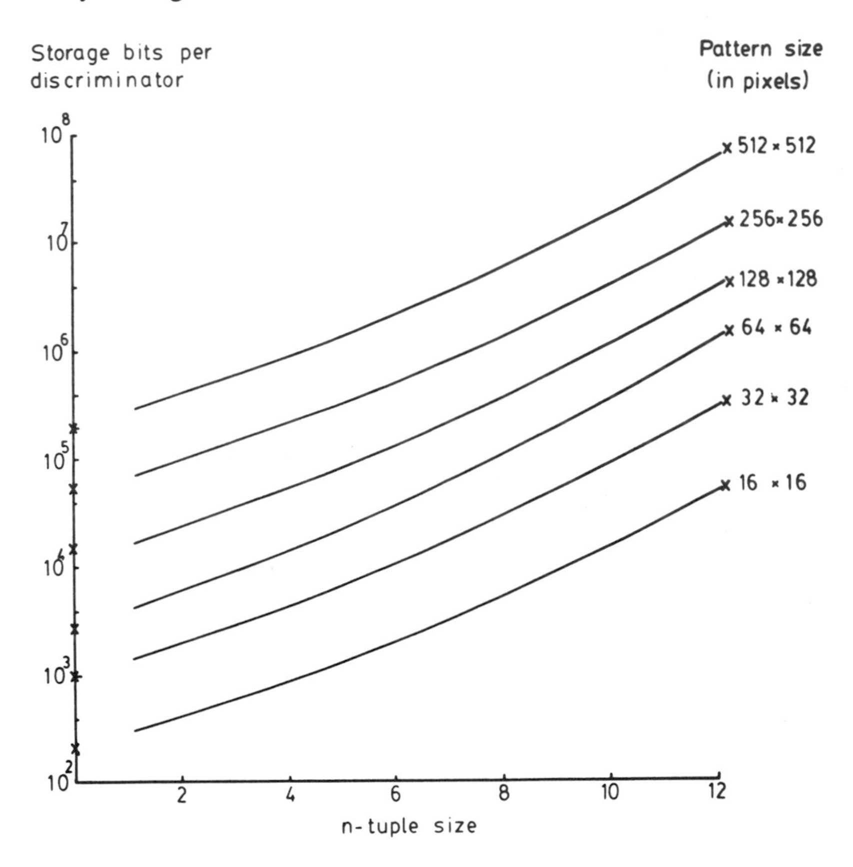

Figure 10. Storage requirement in an n-tuple recognizer

Recognition of Piece-parts

The silhouettes of two different keys are shown in Fig. 11. Each key was allowed to take up any position in a frame with an aspect ratio of 2 to 1, provided it was completely contained within the frame. The maximum dimension of this frame was approximately equal to the greatest dimension of the key. Two discriminators were trained, one for each key, on digitized patterns taken for the frame at a resolution of 32 × 16 pixels. The discriminators each comprised 64 memory elements addressed by random 8-tuple samples, and the keys were removed and reinserted into the frame after each training operation, in order to obtain position independent recognition within the confines of the frame. The system

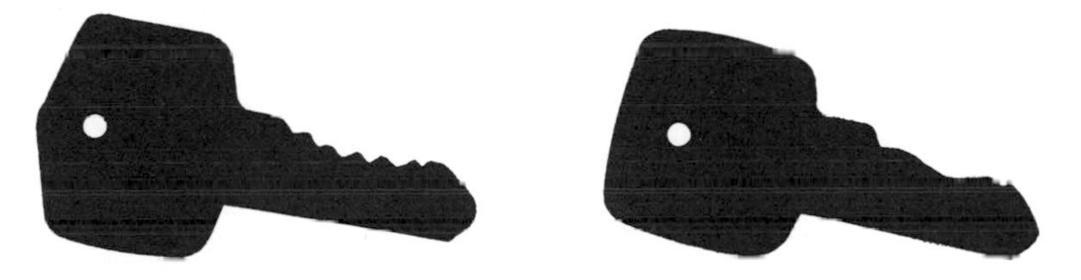

Figure 11. Two keys

was beginning to distinguish between the patterns after 20 trainings. After 50 training cycles the system, evaluated over a series of 20 tests, was able to distinguish, without error, between the two keys. The salient points from this example are:

1. The shapes on inspection are very similar. The human may take a few seconds to notice the difference by comparison.
2. No shape description or comparison criterion has to be supplied to the designer or the machine.
3. The system develops a recognition capability independent of position (within the confines of the frame) without requiring any increase in memory in the processor.

Unconstrained Piece-parts

In the previous problem the keys, although allowed to take up any position in the frame, were in effect constrained to less than 30° of rotational freedom per quadrant. If parts can be observed in any orientation during training, the generalization set will be considerably larger, though not necessarily resulting in overgeneralization.

If the discriminator shown in Fig. 8 is trained on a rotating bar pattern.

```
. x .    . . x    . . .    x . .
. x .    . x .    x x x    . x .
. x .    x . .    . . .    . . x
```

the generalization set would only contain 32 patterns out of a universe of 512, even though almost every point on the input space has had values of both 1 and 0 at some stage during training.

Training data can, of course, comprise digital pattern of objects in any orientation. The generalization sets associated with such patterns would be larger than with constrained data, and reject and error conditions more likely to occur. Nevertheless useful pattern recognition can still be achieved.

A four-discriminator classifier was trained to recognize silhouettes of bolts in any orientation. The images are shown in Fig. 12, and the bolts were allowed to rest in any orientation on a flat surface. A window of 16 × 16 pixel resolution was shrunk onto the digitized image. The objects were displayed on a rotating turntable and camera shots taken at random in order to obtain images in any orientation.

Complete separation of the images was possible after training on 25 shots of each object. The responses of the discriminators to type (a) patterns are illustrated in Fig. 13.

With unconstrained training data the generalization sets are larger, and discrimination between highly similar piece-parts will be impaired. The use of orientation compensating mappings has been proposed [6] and shows significant improvements in both recognition performance and the confidence level of the results, for unconstrained piece-parts.

Identification of Highly Constrained Piece-parts

In certain areas it is possible to constrain piece-parts to fixed positions and orientations manually. If, for example, the outgoings from a part

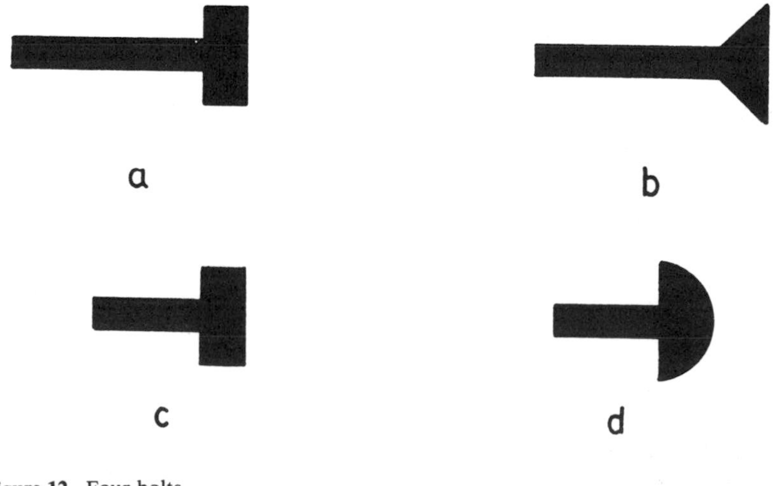

Figure 12. Four bolts

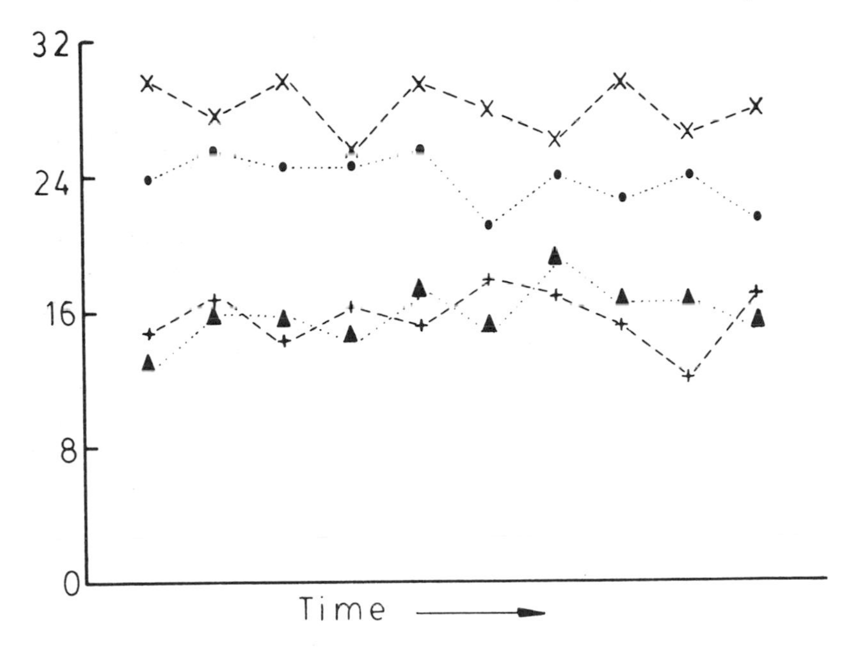

Figure 13. Responses of discriminators to rotated versions of bolt (a).

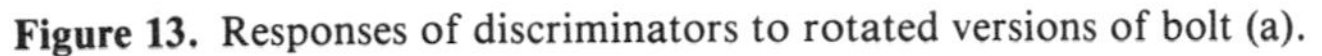

× response of disc (a)
● response of disc (b)
+ response of disc (c)
▲ response of disc (d)

store are to be monitored and an item can only be removed after having been identified and recorded by a pattern recognizer, the parts can be precisely located in a viewing jig. Variation between different images obtained from a given piece-part viewed by a camera arise mainly from electronic noise and the effects of spatial quantization, which can be reduced to acceptable, if not negligible, levels. In practice, a situation is approached whereby one pattern defines a data class and—in the case of a part store—possibly tens of thousands of piece-parts (data classes) have to be recognized.

The assignment of a single discriminator to detect each piece-part becomes impracticable under these conditions. The amount of storage would be far greater than a direct library store of the patterns (the latter, however, would be restricted to serial implementation) and the generalization properties of the networks would not be exploited. The feasibility of using a discriminator in a highly deterministic pattern environment, to detect more than one data category, has therefore been examined.

An example of such a system, a "2 in N" classifier, contains N discriminators, and each discriminator is trained on N−1 categories in such a way that no two discriminators are trained on identical sets of categories, but each category of data occurs in the training sets of two discriminators. The training specification for a four-discriminator classifier is summarized in Table 1.

It can be shown that N discriminators can accommodate $(N^2 - N)/2$ data categories. Hence a 34-discriminator classifier required to recognize fuzzy alphanumeric characters can be configured as a "2 in N" classifier to detect 630 highly constrained piece-parts.

However, as any given discriminator is now trained on pattern categories which are in no way related, the generalization associated with these patterns becomes arbitrary. If in operation the data are highly constrained and well defined and such that one is only required to

Table 1
Training specification for a 4-discriminator "2 in N" classifier

TRAINING SPECIFICATION				
Discriminator	1	2	3	4
Training Categories	A B C	A E D	F B D	F E C

CATEGORY/DISCRIMINATOR ASSIGNMENT	
Category	Detected by discriminators
A	1 and 2
B	1 and 3
C	1 and 4
D	2 and 3
E	2 and 4
F	3 and 4

recognize the training data, any generalization between training patterns may degrade the overall performance. (The amount of generalization can be controlled by the n-tuple size which in itself determines the storage requirements.) Nevertheless, any patterns the system fails to classify can be detected as rejections, as opposed to errors, since in a fully specified system a training pattern will always produce a maximum response from the two discriminators it was assigned to during training. If other discriminators produce maximum responses because of generalization between their training sets, a decision cannot be made on the basis of two discriminators giving maximum responses, hence the part can be assigned as unclassifiable by the system. It will not be given the incorrect classification.

Pattern Detection in Unspecifiable Data

The final application to be considered concerns pattern recognition where the relationship between data and classification is unknown or unspecifiable. An adaptive classifier is trained on sets of patterns of known descriptions and assessed on a test set which was not referenced during training. A classification which is significantly better than chance implies that the data do contain information which is characteristic of its particular classification.

Pattern recognition was applied to a data set of 600 medical files relating to patients admitted to hospital with abdominal pain symptoms of the following diseases or conditions:

Appendicitis
Diverticulitis
Cholecystitis
Small bowel obstruction
Perforated ulcer
Pancrealitis
Non-specific abdominal pain

Each file contained information about the onset and characteristics of the abdominal pain, previous patient history, and details of a physical examination.

The information from each file was encoded into fixed format binary patterns and a classifier comprising 7 discriminators used to partition the data set. The input pattern size was 192 bits and the n-tuple size was set at 4. Each discriminator was trained on 20 patterns associated with one of the disease categories, and the remaining patterns (460) used to assess the performance of the classifier. The results are summarized in Fig. 14 and show that 95% correct partition of the data could be achieved, taking

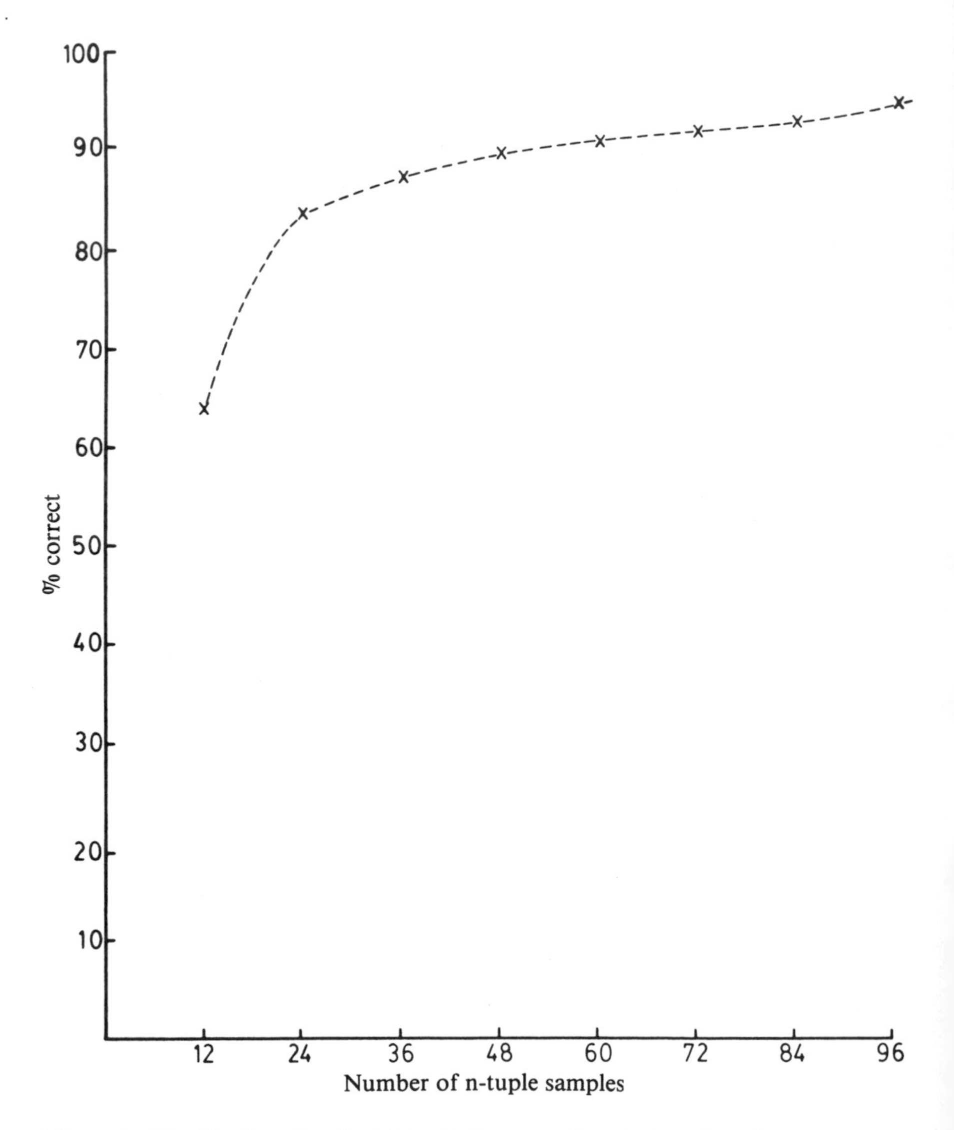

Figure 14. Classification of medical data. Performance % against number of 4-tuple samples

96 random 4-tuple samples of the input space—a total storage requirement of $96 \times 2^4 \times 7 = 10{,}752$ bits. (The classifier could be accommodated within one commercially available 16K bit RAM.) The results are significantly better than chance, where a 14% success rate would be expected.

This pattern detection approach can be applied to many other forms of data, and may be especially appropriate to applications generally regarded as requiring human experience. The monitoring of machines and the process control of a plant are but two examples in the field of industrial automation.

Summary

Networks of memory elements provide means of identifying a wide range of binary patterns. A formal description of the data does not have to be made available, as the recognition mechanism does not rely on any established method of analysis of the data it is classifying. The recognition functions are derived from a known representative set of patterns of each data class. The memory elements in the discriminators do not store patterns (they assume the role of adaptive logic circuits which, after training, detect, for a given category, allowable n-tuple subpatterns on the input space). Therefore the storage per discriminator and operating time are independent of the number of training patterns per category.

The technique can be applied to pattern detection problems, hence the opportunity for establishing hitherto unknown relationships within a data base can be exploited.

Finally, the physical implementation of a pattern recognizer based on networks of memory elements is very flexible. Where slow pattern processing is acceptable (of the order of seconds) a serial simulation on a conventional computer or microprocessor based system can be considered. Hardware versions using RAMs or ROMs will provide faster processing, the operating speed being dependent on the degree to which the system has been structured for parallel processing.

Acknowledgements

The author wishes to thank Mr M. Arain of Brunel University in respect of the medical pattern recognition data.

The financial support from the United Kingdom Science Research Council (WISARD Project) is also acknowledged.

References

1. T. J. Stonham (1976): A classification system for alphanumeric characters based on learning network techniques. *Digital Processes,* vol. 2, p. 321.
2. W. W. Bledsoe and I. Browning (1959): Pattern recognition and reading by machine. *Proc. Eastern Joint Comp. Conf.,* p. 225.
3. I. Aleksander (1970): Microcircuit learning nets—Hamming distance behaviour. *Electronic Letters,* vol. 6, p. 134.
4. T. J. Stonham (1977): Improved Hamming distance analysis for digital learning networks. *Electronic Letters,* vol. 6, p. 155.
5. I. Aleksander and T. J. Stonham (1979): A guide to pattern recognition using random access memories. *Computers & Digital Techniques,* vol. 2, p. 29.
6. L. Gupta (1980): Orientation independent pattern recognition with networks of memory elements. M. Sc. dissertation, Brunel University, U.K.

Chapter 10

Computer Vision Systems for Industry: Comparisons

I. ALEKSANDER
T. J. STONHAM
B. A. WILKIE

Department of Electrical Engineering & Electronics
Brunel University, England

Abstract This article is written for those who are not familiar with the problems of automatic, computer-based pattern recognition. It surveys known methods in the light of opportunities offered by silicon chip technology.
The article also discusses some of the design decisions made in the creation of WISARD*, a fast pattern recognition computer built at Brunel University. Its structure has been optimized for silicon chip implementation.

Is it Difficult for Computers to Recognize Patterns?

As one scans the words of this article and translates the symbols into meaning, the eyes and brain are busy doing an intricate data processing task. This process is called pattern recognition.

When human beings do it, it seems easy and natural. Maybe that is because human beings do it almost all the time: not only when reading or looking at pictures, but also when driving a car, playing tennis, saying hello to the cat, or assembling a watch. All these tasks involve an element of pattern recognition.

So, it is hardly surprising that in the mechanization of production processes there is a need for mechanized, or computerized, pattern recognition schemes. This is not a new revelation: indeed, attempts at computer-based pattern recognition date almost as far back in technology history as the computer itself.

As a result of this it may appear curious that pattern recognition should be regarded as a difficult problem in an age when microprocessors do everything from playing Beethoven's Fifth Sympho-

*WISARD: Mnemonic for *W*ilkie, *I*gor and *S*tonham's *R*ecognition *D*evice, a computer with a novel architecture, built with the support of the U.K. Science and Engineering Research Council, to whom thanks are due.

ny on one's wristwatch to cooking a steak in an automated microwave oven. However, the difficulties are considerable and have appeared insurmountable at times.

Before looking at some of the difficulties and the solutions adopted by the team at Brunel, we shall describe some real areas of need encountered by the authors in their discussions with potential users of pattern recognition machines.

Who Needs Pattern Recognition?

Since the late 1950's the driving force for automatic pattern recognition was not commercial but military. Could one process a picture of the jungle taken from a helicopter hovering over Korea and work out whether there were any hidden tanks or antiaircraft guns in the image? Then, in the early 1960's, postal engineers all over the world thought it would be nice to have automatic reading machines which would decipher postal codes even if written in an untidy scrawl. Strangely enough, neither of these applications had enough force to bring pattern recognition to life in a commercial sense, perhaps because most methodologies did not work fast enough and cheaply enough with the available technology.

It now seems clear that there are two major areas where a sophisticated electronic eye and a silicon-chip brain are needed: the first of these being the automatic production line. The sorting of piece parts and the control of their quality is central to the automation process. The second broad area is simply that of automatic reading: postal codes if needs be, but more pertinently, data on checks, census forms, supermarket labels, V.A.T. forms, library tickets, passports, banker's cards, and so on. On the whole, it seems that wherever one finds human drudgery, there is a need for automatic pattern recognition.

It needs to be said that here one is not trading in redundancy for all production-line workers and clerks. As in the case with all really worthwhile instances of automation, the cost effectiveness comes from the fact that new technology promises an extension and development of what can be done with human effort. As an example, some manufacturers of roller bearings can only afford to examine their product on a sample basis. However, it has been shown that serious failures sometimes do not happen as a result of faults distributed in a neat statistical way that can be spotted in a sample. It's the crack in the odd item that can be disastrous. Still, the quality control engineer has to weigh this low probability against the cost (and operator boredom) of inspecting every part.

The Public Records Office (as an example of the second category) has lorry loads of decaying documents. It would take a vast army of typists to enter the information they contain into computers. This is a clear case of automatic reading where the eventual pay-off has a human rather than a direct, commercial value. In a similar vein, the reading of print in all fonts as an aid for the blind has undeniable human value. This has to be cheap to be used at all.

In medicine as a whole there is a vast set of needs for automatic vision systems ranging from screening (e.g., cervical smears, X-ray plates) to the simple problem of sorting medicament packages through reading codes that are also readable for humans. (At present, color coding and bar coding is used, but this leaves a possible credibility gap since the machine and the human read different data).

On the other hand, the automatic verification of signatures on checks is a highly cost-effective application of the second type. Most clearing banks need to refer checks to the originating branch for verification that the signature bears some resemblance to that of the client of that branch. The major source of fraud which necessitates these massive transfers of paper is the instance where, after stealing a check book, the thief signs the customer's name on the check in his own handwriting, given that the owner's name is clearly written in the bottom right-hand corner of the check. Expert forgers who can sign at speed are few and do not cause much concern. Thus, an automatic verification system which checks the signature against a set of check digits on the check would save the vast sums spent in just moving bits of paper between banks.

Another application is security procedures where the recognition of personal characteristics other than signatures come into view. For example, it would be useful to check faces or fingerprints as means of personal identification. Indeed, in the security area, the state of occupation of an entire building could be checked by a good vision system shared among many television cameras.

Under the buzzword "Robotics" there is the obvious area of vision for industrial automation processes including giving sight to manipulator-arm robots. These find their way into production processes ranging from the welding of car bodies to the grading of apples. In fact, wherever there is a conveyor-belt type of production process there is an application for a vision system comprising a good vision transducer backed by a fast and reliable pattern classifier.

In fact, it is the transducer end of vision which defines the computational difficulty in vision and recognition. We shall therefore define a specification which could be seen as a reasonable upper bound that covers a large number of applications. Clearly, having established such a limit, one can work down from it in cost and systems complexity, which is better than starting with a methodology which lets one down when applied to real problems. The latter has been the case with so many pattern recognition systems in the past.

What the Automatic Eye Needs to See

Probably the cheapest and most reliable vision transducer is the television camera. Also, since some of the tasks for vision automation fall into the category of that which can be distinguished by the human eye on a TV screen, the resolution and accuracy of a TV image appears reasonable for a general-purpose machine. This would clearly include not only vidicon systems as used at present in TV but also solid-state devices such as photodiode arrays and charge-coupled devices.

However, the point of this discussion is to calculate the amount of data that one should aim to process in a general-purpose vision system. Working with the nearest powers of 2 and square. black-and-white images, one takes 512 (which is 2^9, the nearest value to the 625 line resolved by a household TV set) to divide the image into 512 × 512 dots called "pixels." The number of intensity levels between black and white which gives a visually smooth image is 16 (i.e., 2^4, which can be represented by 4 bits of information). One says that such a system has 16 "grey levels."

In any picture processing task using a TV image, one must store at least one frame of the image (a TV camera produces 25 such frames in one second). Such a frame contains 512 × 512 × 4 = 1,048,576 bits. In conventional computer jargon, since information is stored in words (i.e., packets of bits) this requires 131,072 words in a small 8-bit word computer such as a Pet or Apple, or 32,768 words in a larger 32-bit computer such as a DEC VAX 11/750.

Clearly this is a heavy demand (impossible for the Apple or Pet) before any intelligent processing takes place.

In the Brunel WISARD system a special store has been built to hold one TV frame. The store holds 262,144 bits, which is sufficient to store 512 × 512 pixels, two grey-level images (i.e., black or white). This decision is based on the authors' practical experience that at this sort of accuracy obtained from the numbered dots used, the two-level image is sufficient. The system either decides automatically or under operator control what the division point between black and white shall be. Also, one can trade off resolution for grey levels. For example, 16 grey levels may be obtained if one deploys 256 × 256 pixels over the image. WISARD also contains facilities for storing window-like parts of the entire image.

What the Automatic Brain Needs to Compute and How

The details of pattern recognition theory and practice may be found in several good books, for example Ullman [1], Batchelor [2], and Fu [3]. Here we merely aim to give a thumbnail sketch of major trends and their match with what is technologically feasible. One should be aware, however, that this is a field in which papers have been published since the mid-1950s. In 1967 Rutovitz [4] wrote that 1200 papers had been written on the subject, while Ullman 1 lists a selection of 468 key papers, probably about 10% of those available at the time.

However, this level of activity was triggered by the rapid development of fast, gigantic serial digital computers and made unreal assumptions about the speed with which such machines can handle data. In a practical sense, there is very little in this literature to help one with the needs discussed earlier.

Before discussing some of these methodologies we shall define some terms.

Pattern: An array of numerical intensity values representing an image "seen" by the transducer.

Current Pattern: The pattern currently "seen" by the systems assumed to be stored in readiness for processing.

Pattern Class: The set of patterns that is to be recognized in the same way. For example, the recognition of hand printed letters involves 26 pattern classes.

Learning Systems: A system which has the flexibility to form part of its own recognition strategy, from being given patterns whose originating pattern class is known and also given to the system.

Training Set: That set of patterns whose classification is known, and which is given to a learning system during the strategy formulation phase of its operation.

Learning Phase: The strategy formulation period in a learning system.

Test Phase: That operational period during which a learned strategy is being tested.

Features: A pre-supposed set of defining characteristics of a set of patterns. For example, the hand printed letter A may be described as three features consisting of two almost upright lines meeting towards the top of the picture and an almost horizontal line joining the two near their middles.

Pre-Processing: An information reduction operation on an image which standardizes the image in some way. Examples are "thinning" or "skeletonizing" of line-like images, or the extraction of major radii from blob-like images. This is a vast subject and thought to be beyond the scope of this article. Nevertheless it should be realized that pre-processing is highly problem-dependent. It is for this reason that no pre-processing is included in the WISARD system. However, general-purpose parallel machines exist which could be used as pre-processors that in some cases could be limited to WISARD [see Wilson (1980) and Duff (1978)].

The major methodologies we shall discuss here go under the names:

(a) Mask and Template Matching
(b) Discriminant Functions
(c) Logical n-tuple methods

Although this leaves out many other trends and strands, it is felt that the computational properties of those missed out are similar to the ones mentioned. The thrust of the discussion is aimed to show that (c) is best suited to novel silicon-chip architectures and for this reason was chosen as the design method for the "brain" of WISARD.

a) *Mask and Template Matching*

This scheme relies on storing a set of representative prototypes from each pattern class. For example, if well-printed and carefully positioned letters are to be recognized, early schemes stored one prototype for each of the 26 pattern classes. Assuming that black-and-white images only are considered, an unknown image is compared point-by-point with each of the stored prototypes scanning for points which are the same. The unknown pattern is assumed to be the class giving the highest score.

Extensions of the scheme can operate with grey-level images and masks built up by learning where probabilities of occurrence of each point occurring during the training phase are stored. Statistical methods based on such probabilities are widely reported in the literature. Many other methodologies have been developed on the basis of mask matching, the best known perhaps being the "nearest neighbor" method where it is usual to store many patterns from the training set and, during the test phase, to measure the difference (or "distance") of the unknown pattern from all the stored patterns, assigning its classification according to what is its "nearest neighbor."

Computationally this scheme may be taxing as storage, while being fairly light on calculation requirements. The storage demands grow with the diversity of patterns within each pattern class. For example, taking the 512 × 512 binary image used to detect horizontal lines from vertical lines of, say 12 pixels in width, the system would have the following specification:

Storage: Two sets of templates are required, each containing about 500 templates (one for each position of the line). That is $2 \times 500 \times 512 \times 512 \simeq 25 \times 10^6$ bits.

This would defeat most computer installations if the storage were to be directly accessed.

Calculation Time: For each new pattern that has to be recognized one has to carry out

$n \times 512 \times 512$ *simple* comparisons of the type

(n is the number of templates):

* load a word from the image into the accumulator
* "exclusive" or "it" with a word from the template
* right shift x times and count the number of 1s in the accumulator (x is the number of bits in the word).

So, to a rough approximation, the decision time for an unknown pattern is

$$\frac{n \times 512 \times 512(1 + x)}{x} \simeq n \times 512 \times 512$$

computation cycles.

This for a computing cycle of, say (optimistically) 1 μs, would take seven or eight minutes to recognize an image, given a conventional machine architecture. If more sophisticated comparisons are used or if more grey levels are necessary, the figures worsen sharply.

Application. It is clear that mask matching is restricted to low-resolution images with a low diversity in each pattern class. This makes it unsuitable for almost all except the most trivial applications.

Discriminant Functions

This method attempts to overcome the problem of having to store many templates for each pattern class in the following way. A numerical value called a weight is associated with each pixel. So if there are k pixels the weights are numbered:

$$w_1, w_2, w_3, \ldots, w_k$$

In the first instance a discriminant function acts as a device which divides its input patterns into two pattern classes only (called a dichotomiser). It does this by using the intensity values $i_1 \ldots i_k$ of the pixels to perform the following calculation: $S = \sum_j i_j \times w_j$

This is known as the "discriminant function." Then, if S is greater or equal to some other parameter G it assigns the pattern to class A (say), whereas otherwise it assigns it to class B.

The crux of this system is to develop the correct values of the weight w_1 to w_k and the "threshhold" T on the basis of a training set. Much has been written on methods for doing this by means of incremental adjustments; however, the sad fact is that in some cases no discriminant function exists to distinguish between two classes. A good example of this is the horizontal/vertical discriminator discussed under (a) above. To demonstrate this one can reduce the image to 2 × 2 binary pixels. Then the horizontal images are:

$$\text{H-class: } \begin{matrix}11\\00\end{matrix}, \begin{matrix}00\\11\end{matrix} \quad \text{V-class: } \begin{matrix}10\\10\end{matrix}, \begin{matrix}01\\01\end{matrix}$$

Then, numbering the image

$$\begin{matrix} i_1 & i_2 \\ i_3 & i_4 \end{matrix}$$

it is relatively simple to show that the following conditions cannot be satisfied with real weights:

$$(w_1 + w_2) > (w_1 + w_3)$$
$$(w_1 + w_2) > (w_2 + w_4)$$
$$(w_3 + w_4) > (w_1 + w_3)$$
$$(w_3 + w_4) > (w_2 + w_4)$$

Extensions of this scheme employ several discriminators simultaneously to overcome situations such as the above. However, in the horizontal/vertical example the method would effectively degenerate to having one template per vertical line.

Computationally the system ostensibly needs less storage than mask matching, since only the weights need to be stored; however, the product and addition calculations take considerably longer storage. Also, the training phase may be extremely long and unproductive. Therefore, given that weights can be found, the specification for the TV-compatible system would be:

Storage: 512×512 bytes (for integer arithmetic) per class pair, which, for one class-pair and eight-bit bytes, comes to
$\simeq 2 \times 10^6$ bits

This compares favorably with mask matching.

Calculation time: allowing 10 μs for multiplication and 3 μs per addition and temporary storage overheads, the TV system ought to be capable of providing a decision in

$$13 \times 10^{-6} \times 512 \times 512$$

$$\simeq 3 \text{ seconds.}$$

Although this compares favorably with mask matching, it falls short of most industrial robotics task where four images per second is a standard. Clearly, recognizing images at TV frame speeds (25/second) seems even further out of reach.

Applicability: Although discriminant functions appear to solve some of the storage and timing overheads found in mask matching, the uncertainty of whether the methodology will work at all with any particular problem, and the remaining storage and computational overheads for conventional architectures have limited the practical applicability of the technique.

c) Logical n-tuple methods

Most of the above methods take little account of joint occurrences among pixels. Indeed, these joint occurrences define the shape of an image. Particularly in industrial robotics where one is generally dealing with solid objects defined by their shape, it is this joint occurrence of events which enables one to identify the shape themselves. It is the power of this method that caused its adaption into the WISARD system.

An n-tuple is a group of pixels which is processed as a simple entity. Thus, for example, a 512×512 pixel image is processed as

$$\frac{512 \times 512}{n}, \text{ n-pixel chunks.}$$

For well-defined sets of images (such as printed numbers on checks) these groupings can be predefined and are called "features." Here we concentrate on more difficult cases where such features cannot be easily predefined. The latter is far more common than the former.

In such cases it is usual to pick the n pixels that form an n-tuple at random in the first instance. Ways of de-randomizing such choices may be found in the literature. The details of the operation of n-tuple systems may be found in Aleksander and Stonhams. (1979). Here we merely present the skeleton of the method so that it may be compared with others. A complete bibliography of the Brunel team's work on n-tuple pattern recognition may be found in Appendix 1.

Operation: The image is partitioned into its n-tuples, each n-tuple being treated as a binary word. So, for example, if n = 8 and 16 (i.e., 2^4) grey levels are used, the bit-length of the n-tuple word is 32. This can encode all possible subpatterns that might occur within the n-tuple. The crux of the n-tuple method is that the n-tuple is used to address store, as can be explained by the following example which, in fact, uses some of the parameters employed in WISARD. Taking a 512 × 512 binary image and n = 8‡, there are 32,768 8-tuples for the image. The way in which they are connected is irrelevant at present. Each 8-tuple addresses its own 256 words of store. We call these D-words or discriminator words. Each D-word has as many bits in it as there are pattern classes.

Whenever a pattern is present at the input each 8-tuple addresses (or selects) just one of its 256 D-words. During testing one totals the number of ls addressed in each bit position and recognizes the pattern as the class which gives the biggest score. Again the reader is referred to the literature for a clarification of this addressing scheme.

Computationally, although the scheme appears to require more storage than template matching this is not true. Also, it needs virtually no calculated overheads and is ideally suited to parallel or semi-parallel (i.e., WISARD) organization. We can see some of this

‡Experience has shown that n = 8 is more than adequate for most applications.

by following through the vertical/horizontal examples discussed earlier.

Storage: Leaving n undetermined for the moment, we note that for pattern classes the number of bits required is:

$$(\text{number of classes}) \times \frac{(\text{size of image})}{n} \times 2^n$$

$$= 2 \times \frac{512 \times 512}{n} \times 2^n$$

$$\simeq .5 \times 10^6 \quad \frac{2^n}{n}.$$

The central questions in all problems solved by using the n-tuple method is "what is a suitable value of n?" Often this can only be answered by experimentation, while in this case it may be shown (by appealing to the case where the n-tuples are placed in either horizontal or vertical lines more than 12 pixels apart) that n = 2 is sufficient. It may also be shown that n = 2 will deal with recognition of many vertical and horizontal lines as well as the estimation of whether a line is more horizontal than vertical. It may further be shown that a random connection with n = 2 will suffice. Hence, evaluating the last formula the resulting storage is:

$$10^6 \text{ bits}$$

which is $\frac{1}{25}$ th of that required with mask matching.

Calculation Time: With a conventional 16 bit (say) architecture, the computation time is mainly one of storage access, once per n-tuple per pattern class. Taking 1 μs (pessimistically) per access we have:

$$\frac{512 \times 512}{2} \times 2 \times 10^{-6} \text{ seconds}$$

$$\simeq \frac{1}{4} \text{ second}$$

Clearly this is an improvement on what was discussed before; nevertheless it is still too high in some realistic situations where the number of pattern classes may be high.

Applicability: It appears that if practical considerations are brought into play, the n-tuple scheme is the most flexible and the fastest among the available systems. It appears to have fewer performance drawbacks than other systems; it is for these reasons that it was chosen for the WISARD and other similar systems. However, the speed problem with most of these schemes is such that it is vital that one should consider using special architectures based on advancing silicon chip technology in order to see whether speed barriers may be broken.

Special Architectures

Looking very briefly at the three methodological categories, we see that they all could be made more efficient by means of special architectures. Mask matching, for example, could be based on a parallel system of registers and comparators. However, it may be shown that the performance of a mask matching system is equivalent to an n-tuple system with n = 1, and therefore not only limited but lacking in generality. Therefore it would appear unwise to launch into the design of such special architectures with mask-matching methodology in mind.

Discriminant function methods rely heavily on calculation, and one could envisage an array system where each cell performs the necessary weight-multiplication calculation. In fact, such architectures exist (Gostick [6] and Duff [7]) and the research may be worth doing. However, the performance uncertainties of the scheme mitigate against optimism.

The n-tuple scheme therefore remains not only as a good way of organizing a conventional architecture but also a good candidate for special architectures.

To achieve speed, one arranges the n-tuples to be addressed in a semi-parallel way, where the degree of parallelism and other characteristics such as window size, value of n etc., are under operator control. WISARD was constructed with these advantages in mind and details of design decisions will be published in due course.

Fixed Implementations

So far, the main concern here has been the structure of learning systems which improve their performance by being "shown" a suitable training set. However, there are many applications where a fixed, pattern recognition task has to be carried out over and over again. The n-tuple method lends itself to a process of reduction which implements either as logic gates, Programmed Logic Arrays or Read-only Memories, the essential parts of the learned logic in a learning system. This is not the place to dwell on details of this type of procedure, except to realize that it is purely a mechanical and easily computable procedure. For example, if the horizontal/vertical problem had been solved with horizontal 2-tuples, with the two elements of the 2-tuple 13 pixels apart, then it can be shown that for the "horizontal" pattern class the memories could be replaced by AND gates, whereas for the "verticals" case these could be replaced by "Exclusive OR" gates. In fact, the designs of such logic is based on the realization that the contents of the store associated with each n-tuple is merely a truth table derived during a learning process. For details of this see Aleksander [8].

Conclusion: Pattern Recognition Comes of Age

A central characteristic of the silicon chip "age" is the remarkable drop it has brought about in the cost of information storage. Even in the last 15 years the cost of storage has dropped by a monetary factor of over 300. In real terms the WISARD recognition store would have cost the equivalent of 10, 4-bedroom houses in London 15 years ago. It now costs as much as one domestic color television set.

It is hardly surprising, therefore, that much theory and methodology in the history of pattern recognition has been held on the wrong economic premises. The point made in this article is that the n-tuple methodolgy promises the means for exploiting this economic advantage. However, this is only the beginning of the story. In the bibliography of Appendix 1 there is much on the way in which one might improve the "intelligence" of machines such as WISARD. In particular, one notes the possibility of closing the loop between the recognition signals in these systems and their input, producing an interesting dynamism. This would enable the system to recognize sequences of events and generate sequences of recognition signals, effectively moving closer to the sequential but powerful way in which humans recognize patterns. The theory of such schemes was pursued before experimentation became economically feasible; therefore it is expected that the practical economic advantages brought about by silicon chips and machines such as WISARD will make interesting experiments possible. This may bring about a novel race of computing machines, more flexible than the conventional computer and better adapted to communication with human beings.

References

1. Ullman, J. R. (1973) Pattern Recognition Techniques, Butterworth.
2. Batchelor, B. (1978) Pattern Recognition: Ideas in Practice.
3. Fu, K. S. (1976) Digital Pattern Recognition, Springer Verlag.
4. Rutovitz, D. (1966) "Pattern Recognition" *J. Roy Stat Soc.*, Series B/4, p. 504.
5. Aleksander, I. & Stonham, T. J. "A Guide to Pattern Recognition Using RAM's", *IEE J. Dig. Sys & Computers*, Vol. 2, No. 1 (1979(+)).
6. Gostick, R. W. ICL Tech fair, 1979, Vol. 1, No. 2, pp. 116-135.
7. Duff, K. J. B. Parallel Processing Techniques, in Batchelor (1978) (see above).
8. Aleksander, I. (1978) Pattern Recognition with Memory Networks, in Batchelor (1978) (see above).

APPENDIX 1

Complete bibliography on Adaptive Pattern Recognition by the Brunel Team

ALEKSANDER, I.
Fused Adaptive Circuit which Learns by Example, Electronics Letters, August 1965.

ALEKSANDER, I.
Design of Universal Logic Circuits, Electronics Letters, August 1966.

ALEKSANDER, I., NOBLE, D. J., ALBROW, R. C.
A Universally Adaptable Monolithic Module, Electronic Communicator, July-August 1967.

ALEKSANDER, I., ALBROW, R. C.
Adaptive Logic Circuits, Computer Journal, May 1968.

ALEKSANDER, I., ALBROW, R. C.
Pattern recognition with Adaptive Logic Elements in "Pattern Recognition," IEE 1968.

ALEKSANDER, I., ALBROW, R. C.
Microcircuit Learning Nets: Some test with hand-written numerals, Electronic Letters, 1968, p. 408.

ALEKSANDER, I., MAMDANI, E. H.
Microcircuit Learning Nets: Improved recognition by means of Pattern feedback, Electronics Letters, 1968, p. 425.

ALEKSANDER, I.
Brain cell to microcircuit, Electronics & Power, *16*, 48–51, 1970.

ALEKSANDER, I.
Some psychological properties of digital learning nets, Int. J. Man-Machine Studies, *2*, 189–212, 1970.

ALEKSANDER, I.
Microcircuit learning nets: Hamming distance behaviour, Electronics Letters, 6, 134, 1970.

ALEKSANDER, I., FAIRHURST, M. C.
Pattern Learning in humans and electronic learning nets, Electronics Letters, 6, 318, 1970.

ALEKSANDER, I., FAIRHURST, M. C.
Natural Clustering in Digital Learning Nets, Electronics Letters, *7*, 724 1971.

ALEKSANDER, I., FAIRHURST, M. C.
An Automation with Brain-like properties, Kybernetes, *1*, 18, 1971.

ALEKSANDER, I.
Electronics for intelligent machines, New Scientist, *49*, 554, 1971.

ALEKSANDER, I.
Artificial Intelligence and All That, Wireless World, October 1971.

ALEKSANDER, I., FAIRHURST, M. C.
Dynamics of the Perception of Patterns in Random Learning Nets, in: Machine Perception of Patterns and Pictures, Phys. Soc. London, 1972.

ALEKSANDER, I.
Action-Oriented Learning Networks, Kybernetes 4, 39–44, 1975.

ALEKSANDER, I., STONHAM, T. J., et al.
Classification of Mass Spectra Using Adaptive Digital Learning Networks Analyt. Chem. *47*, 11, 1817–1824, 1975.

ALEKSANDER, I.
Pattern Recognition with Networks of Memory Elements, B. Batchelor. Plenum Publications, in 'Pattern Recognition: Ideas in Practice,' London 1976.

ALEKSANDER, I., STONHAM, T. J.
Guide to Pattern Recognition using Random-Access Memories, Computers & Digital Techniques, Feb. 1979, Vol. 2, No. 1.

ALEKSANDER, I.
Intelligent memories and the Silicon chip, Electronics and Power Jour. April 1980.

WILSON, M. J. D.
Artificial Perception in Adaptive Arrays, IEE Trans. on Systems, Man and Cybernetics, Vol. X, No. 1, 1980, pp. 25–32.

STONHAM, T. J.
Improved Hamming Distance Analysis for Digital Learning Networks, Electronics Letters, Vol. 6, p. 155, 1977.

STONHAM, T. J.
Automatic Classification of Mass Spectra, Pattern Recognition, Vol. 7, p. 235, 1975.

STONHAM, T. J., ALEKSANDER, I.
Optimisation of Digital Learning Networks when applied to Pattern Recognition of Mass Spectra, Electronics Letters, Vol. 10., p. 301, 1974.

STONHAM, T. J., FAIRHURST, M. C.
A Classification System for Alpha-numeric Characters Based on Learning Network Techniques, Digital Processes, Vol. 2, p. 321, 1976.

STONHAM, T. J., ALEKSANDER, I.
Automatic Classification of Mass Spectra by Means of Digital Learning Networks, Electronics Letters, Vol. 9, p. 391, 1973.

STONHAM, T. J., ENAYAT, A.
A Pattern Recognition Method for the Interpretation of Mass Spectra of Mixtures of Compounds, Internal Report, University of Kent, 1977.

ALEKSANDER, I., AL-BANDAR, S.
Adaptively Designed Test Logic for Digital Circuits, Electronics Letters, Vol. 13, p. 466, 1977.

STONHAM, T. J., FAIRHURST, M. C.
Mechanisms for Variable Rejection Rate in a N-tuple Pattern Classifer, Internal Report, Unviversity of Kent, 1976.

Chapter 11

Memory Networks for Practical Vision Systems: Design Calculations

I. ALEKSANDER

Department of Electrical Engineering and Electronics
Brunel University, England

Abstract If memory networks are to be used in real robotic applications, the designer needs some basis from which to calculate the size of the modules required and the amount of training that provides optimal results. This paper is concerned with the recognition of solid shapes, and provides design calculations and methods for systems that recognize objects irrespective of their location or orientation. It tackles the problem of the recognition of partially hidden or overlapping objects, and the measurement of position and orientation itself.

Introduction

For most, neural modelling in the design of image recognition systems has a historical beginning in 1943 with the work of McCulloch and Pitts [1] and a quasi-death in 1969 with the condemnation by Minsky and Papert [2] of Rosenblatt's development of such a model into a "perceptron" [3]. The argument in [2] centres on the blindness of such systems to geometrical and topological properties of images for which the calculational powers offered by classical computer structures are better suited. It was also seen that the inferential power of some artificial intelligence (AI) programs offered more to image recognition (or "scene analysis") than the neural modelling approach.

This paper is a report on work which might, in some sense, have appeared to be "heretical" in the late 1960s [4], as it suggested that the neural model could be used to design systems which would benefit from novel solid-state architectures. Indeed, the practical recognition system called WISARD* [5] has been based on the use of 32,000 random access memory (RAM) silicon chip devices as "neurons" in a single-layer "neural net". This system will be addressed indirectly in parts of this paper, while the central aim is to

*Wilkie's, Stonham's and Aleksander's Recognition Device.

provide a new assessment of such nets in image recognition tasks important in robotics. This is a major application area for WISARD, as it can take decisions on 512×512 pixel images at the rate of 5-10 per second. Robotic vision tasks require decisions at this order of speed on rotated and translated versions of objects and sometimes demand the recognition of combinations of objects even if occlusion takes place. The behaviour of such networks in this area will be assessed using a theoretical methodology somewhat different from empirical and problem-orientated methods which are common in this field. The methodology adopted here is to lay a general theoretical foundation for the operation of the system, show its sensitivity to key parameters and then employ an analytically tractable group of simplified shape-analysis problems to illustrate the way in which problem definition leads to the choice of parameters. The paper ends with a description of current work on dynamic versions of such nets, and expectations for the future.

The RAM/Neuron Analogy

Although this has been published before, it is included here in a different format for completeness. The central function of most neural models is the discriminant function:

$$\Sigma_i w_i x_i = \Theta \tag{11-1}$$

where x_i is the ith, binary synaptic input of the neuron and w_i is its associated synaptic "weight", $-1 < w_i < 1$ whereas θ is a threshold such that the neuronal axon fires if $\Sigma_i w_i x_i \geqslant \theta$ and does not otherwise.

If the input consists of N synapses, for instance $1 \leqslant i \leqslant N, i = 1, 2, 3$ etc, equation (11-1) can be written in the more general form:

$$X \overset{W}{\rightarrow} \{0, 1\} \tag{11-2}$$

where X is the set $\{\underline{x}_1, \underline{x}_2, \underline{x}_3 \ldots \underline{x}_2N\}$ of binary patterns and $\underline{W}$ is the weight vector $(w_1, w_2, \ldots w_j \ldots w_N)^T$, whereas $\{0,1\}$ indicate firing, 1 or not, 0. T indicates a vector transposition. A RAM, on the other hand, performs the mapping:

$$X \overset{M}{\rightarrow} \{0, 1\} \tag{11-3}$$

where M is a binary vector of the form ($m_0, m_1 \ldots m_{2N}$-1), which is, in fact, the content of the memory, where the "inputs" of the memory are N "address" terminals, and the output is a single binary terminal. Thus, m_j is a single-bit "word" addressed by the jth pattern from X. One notes the following differences between the model represented by equation (11-2) and that represented by equation (11-3):

1. In the former, the logical function is stored as a vector $\underline{W}$ of N continuous variables while in the latter it is stored in a binary vector M of 2^{N-1} variables.
2. The former can achieve only "linearly separable" functions, which is a number much inferior (but hard to calculate [6]) to all the functions (2^N of them) that the latter can achieve, each being characterized by a distinct $\underline{M}$ vector in the latter case.

None of this so far describes the way that the $\underline{W}$ or $\underline{M}$ values are built up, that is, the way the model "learns". In most cases the procedure from the point of view of a user is much the same. The system is "given" the desired output for specific $\underline{x}_j$ patterns present at the input (from a "training" set). Some machinery needs to be put in motion within the system to bring about the desired mapping. In neural models using $\underline{W}$ this can be quite a complex procedure needing many iterations [6]. In the latter models the bits m_j of M are set directly, with the "desired" response being fed to the RAM through its "data input" terminal.

When "perceptrons" were central to image recognition work it was thought that the differences mentioned in [1] gave the system some of its "intelligence" since fewer examples of the $X \xrightarrow{\underline{W}} \{0,1\}$ mapping need be shown to the system in order for it to "fix" its function. Now, is this "fixing" an arbitrary process which is as likely to worsen performance as to improve it? As will be seen, it is the structure of the network that allows a measure of control over the discriminatory powers of the system, and not so much the generalization effects within the elements of the net. Thus, the total lack of generalization within RAMs is of little consequence; what is important is the way they are deployed in networks and the way such networks are trained.

Technologically, $\underline{W}$ calls for storage of analog or highly multivalued data. This has led some experimenters to build machines in which $\underline{W}$ was implemented with roomfuls of potentiometers or special registers. The RAM system, on the other hand, leads to highly cost-effective systems which can either be implemented directly on conventional machines (leading to one or two decisions per second on a 256-kilobit vector), or be implemented by hardware totally parallel (which could give up to 10^6 decisions per second), or with semi-parallel WISARD-like systems [7, 8] which give tens of decisions per second at a relatively low cost.

Single Layer, Randomly Connected Nets: Parameters and Definitions

R is the number of bits in the image to be recognized.
N is the number of address (synaptic) inputs to each RAM, known as *N-tuple*.
A *discriminator* is a group of K RAMS where K usually is R/N as a discriminator is made to "cover" the image just once, in a one-to-one manner. The connection is made at *random*.
C is the number of classes to be distinguished in a particular problem. This implies the existence of C discriminators, one per class.

Training is the process of presenting the system (at the overall input) with examples of patterns in the ith class and "writing" into the ith discriminator, logical 1s at all of its K "data in" terminals. It is assumed that all the memories are reset to 0 at the start of a training run.

A *response* r_j of the jth discriminator to an arbitrary pattern $\underline{x}_i$ is obtained as follows. First, note that each discriminator generates an integer c_j:

$$0 \leqslant c_j \leqslant K, \text{ for all } j \text{ from 1 to } C. \qquad (11\text{-}4)$$

The response is then quoted as a percentage of K, for instance:

$$r_j = c_j/K \ \%$$

In this connection also define a *response vector* to $\underline{x}_i$:

$$Z(\underline{x}_i) = (r_1, r_2 \ldots r_C)^T \qquad (11\text{-}5)$$

A *decision* is made on the basis of $Z(\underline{x}_i)$, the decision being a singular mapping:

$$Z(x_i) \overset{\delta}{\rightarrow} \{d / d \in \{1, 2 \ldots K\} U \phi\} \qquad (11\text{-}6)$$

where d is the class assigned to the unknown pattern and ϕ is a symbol for a "don't know" decision. Several *decision strategies* may be employed, for example:

1. The *maximum response strategy*, where d is the value of j for the maximum r_j in $Z(\underline{x}_i)$. This is written:

   $$d = JMAX[Z(\underline{x}_i)] \qquad (11\text{-}7)$$

   whereas the actual response of the discriminator in question is written:

   $$f = MAX[Z(\underline{x}_i)] \qquad (11\text{-}8)$$

2. The *confidence conditioned strategy*, where the highest response $Z(\underline{x}_i)$ with d removed exceeds a pre-specified threshold T. This is written:

   $$d - JMAX[Z(\underline{x}_i)] \qquad (11\text{-}9)$$

   if:

   $$MAX[Z(\underline{x}_i)] - MAX[Z(\underline{x}_i) - d] > T_c \qquad (11\text{-}10)$$

   or else $d = \phi$.

3. The *average conditioned strategy*, where the average response must exceed a pre-specified threshold T_a before the decision is noted. This is written:

   $$d = JMAX[Z(\underline{x}_i)] \qquad (11\text{-}11)$$

   if:

   $$AVE[Z(\underline{x}_i)] > T \qquad (11\text{-}12)$$

or else $d = \phi$. There are many others used in specific applications.

The system treats the image as an R-bit vector. Should there be grey levels involved, and should the image consist of p pixels of b bits each:

$$R = p \times b \qquad (11\text{-}13)$$

The above description should be read as being ''notional''. This means that its implementation could either be in the form given, in which case a fast, parallel system would result or both semi-parallel and serial reorganizations of the ''notion'' are possible [7, 8].

Analysis of System Training and Responses

The theory and early understanding of such systems were provided not by those interested in neural modelling but by workers in pattern recognition. This originated with work on N-tuple techniques by Bledsoe and Browning [9] to which much was added in the context of character recognition by Ullman [10]. Some general analysis may be found in work by the author [11, 12], again largely in the area of character recognition.

The objective here is to provide a simplified theoretical background relevant in situations such as robot vision, where the data is originally derived from shapes and variations and provided by rotation translations or distortions of such shapes. Three effects will be pursued:

1. the probable response of a single discriminator as a function of N and training;
2. the probable response of several discriminators as a function of N and training;
3. the probable saturation of the system as a function of N and training.

The equation which is common to all three of the above effects, in the probability of siting a randomly selected N-tuple, is a specific field of Q points in a total field of R points. We call this P(N, Q, R) where:

$$P(N, Q, R) = \frac{Q!}{N!\,(Q-N)!} \div \frac{R!}{N!\,(R-N)!} = \frac{Q!\,(R-N)!}{R!\,(Q-N)!} \quad (11\text{-}14)$$

This holds as long as $1 \leqslant N \leqslant Q$ $\quad N = 1, 2, 3$ etc, whereas $P(N, Q, R) = 0$ when $N > Q$.

Single Discriminator Analysis

For clarity we simplify the analysis to a system that has been trained on only two patterns $\underline{x}_1$ and $\underline{x}_2$ and is tested with a third $\underline{x}_3$. We define the *similarity* between two patterns (say $\underline{x}_1$ and $\underline{x}_2$) as:

$$\underline{s}_{12} = \underline{x}_1 \ominus \underline{x}_2 \quad (11\text{-}15)$$

$\ominus$ is the equivalence operator such that s_i of $\underline{s}_{12} = 1$ if x_i of $\underline{x}_1$ is equal to x_i in $\underline{x}_2$.

We further define the *area of similarity* S_{12} related to $\underline{s}_{12}$ as the number of 1s in $\underline{s}_{12}$. Now it is possible to compute the probability $P(\underline{x}_3)$ of generating an output of 1 from a specific RAM in response to $\underline{x}_3$. This is achieved when its N-tuple is sited either in the area of similarity between $\underline{x}_3$ and $\underline{x}_1$ *or* the area of similarity between $\underline{x}_3$ and $\underline{x}_2$ *but* not both (to avoid double counting). This can be written using equation (11-14) as:

$$P(\underline{x}_3) = P(N, S_{13}, R) + P(N, S_{23}, R) - P(N, \{\underline{s}_{13} \cap \underline{s}_{23}\}, R) \quad (11\text{-}16)$$

Note that x_i in $\{\underline{s}_{13} \cap \underline{s}_{23}\}$ is 1 if x_i is 1 in $\underline{s}_{13}$ *and* x_i is 1 in $\underline{s}_{23}$. It is clear that Boolean algebra applies to the input vectors and it may be shown that the area of $\underline{s}_{13} \cap \underline{s}_{23}$ is the total overlap between s_1, s_2 and s_3. Let this be S_{123}. Hence:

$$P(\underline{x}_3) = \frac{(R-N)!}{R}\left[\frac{S_{13}!}{(S_{13}-N)!} + \frac{S_{23}!}{(S_{23}-N)!} - \frac{S_{123}!}{(S_{123}-N)!}\right] \quad (11\text{-}17)$$

Note that this leads directly to the most likely response of the discriminator (say the jth one):

$r_j = P(\underline{x}_3)$ or, for a better way of relating r_j to $\underline{x}_3$, this may be written:

$$r_j(\underline{x}_3) = P(\underline{x}_3) \tag{11-18}$$

To begin to appreciate the effect of N, first let $N = 1$. It may be shown that equation (11-17) collapses to:

$$r_j(\underline{x}_3) = \frac{\text{Area of } [\underline{x}_3 \cap (\underline{x}_1 \cup \underline{x}_2)]}{R} \tag{11-19}$$

The significance of this is that $r_j(x_3)$ reaches a maximum (for instance 100%) when $\underline{x}_3 = \underline{x}_1 \cup \underline{x}_2$. That is, the discrimination is so poor that the system confuses $\underline{x}_1 \cup \underline{x}_2$ with the individual $\underline{x}_1$ and $\underline{x}_2$ (the latter also giving a 100% response).

Now assume that $\underline{x}_3 = \underline{x}_1 \cup \underline{x}_2$ and that $1 < N \leqslant R$. In computing $r(\underline{x}_1)$ it is noted that $S_{11} = R$, $S_{12} = S_{123}$, hence equation (11-17) gives:

$$r_j(\underline{x}_1) = \frac{(R-N)!}{R!}\left[\frac{R!}{(R-N)!} + \frac{S_{12}!}{(S_{12}-N)!} - \frac{S_{12}!}{(S_{12}-N)!}\right] \tag{11-20}$$

By the same reasoning:

$$r(\underline{x}_2) = 100\% \text{ (irrespective of N)} \tag{11-21}$$

It may be shown using equation (11-17) that $r_j(\underline{x}_3)$ rapidly and monotonically decreases to zero as N increases, certainly having reached it by the time N reaches S_{13} or S_{23}, whichever is the greater.

These principles of increasing discrimination against combinations of training patterns apply clearly to greater numbers of training patterns, whereas, as can be imagined, the equivalents of equation (11-17) become a rather messy series of irreducible terms. The main fact remains that a single discriminator has a discriminatory power that increases sharply and monotonically with N.

Multi-discriminator Analysis

Again only a simple case with two discriminators is considered in order to demonstrate trends which occur in more complex situations. Assume that there are two discriminators giving responses r_1 and r_2 and that each is trained on one pattern, say $\underline{x}_1$ and $\underline{x}_2$ respectively. The system is tested with a third pattern $\underline{x}_3$. Then the responses to the third pattern are in full notation: $r_1(\underline{x}_3)$ and $r_2(\underline{x}_3)$. According to equations (11-7) and (11-8) the decision goes to:

$$d = \text{JMAX}[r_1(\underline{x}_3), r_2(\underline{x}_3)] \tag{11-22}$$

$$f = \text{MAX}[r_1(\underline{x}_3), r_2(\underline{x}_3)] \tag{11-23}$$

Without loss of generality let $d = 1$, $f = r_1(\underline{x}_3)$. It is convenient to redefine the confidence on d in equation (11-10) in the following relative way:

$$\text{CON}(d) = [r_1(\underline{x}_3) - r_2(\underline{x}_3)] \div [r_1(\underline{x}_3)] \tag{11-24}$$

Noting that, from equation (11-17) the responses are:

$$r_1(\underline{x}_3) = \frac{(R-N)!}{R!} \cdot \frac{S_{13}!}{(S_{13}-N)!}, \quad r_2(\underline{x}_3) = \frac{(R-N)!}{R!} \cdot \frac{S_{23}!}{(S_{23}-N)!} \tag{11-25}$$

Then, we have

$$\text{CON}(d) = 1 - \frac{S_{23}(S_{23}-1)(S_{23}-2)\ldots(S_{23}-N+1)}{S_{13}(S_{13}-1)(S_{13}-2)\ldots(S_{13}-N+1)} \tag{11-26}$$

To clarify the role of N, note that generally $N \ll S_{13}$ or S_{23} and S_{13} or $S_{23} \gg 1$. Then, to observe trends it can be said roughly that:

$$\text{CON}(d) \cong 1 - \left(\frac{S_{23}}{S_{13}}\right)^N \tag{11-27}$$

Thus, not only does increasing N improve the discrimination properties of a single discriminator, but also provides for improved discrimination within a multi-discriminator system in an exponential way.

However, recall that the storage costs of the system increase exponentially with N too, and it is also possible to degrade the performance of the system through too high a discriminatory power, hence the choice of N is a compromise between these factors [10,12]. Optimal values of N for the recognition of hand-printed characters are central to [10].

Saturation

Starting with all RAMs at zero, the total number of logical 1s in the memory of a discriminator is a function of the number of training patterns seen by that discriminator. Let the average relative number of 1s in a single RAM be:

$$M(t) = [\text{number of 1s}]/2^N \tag{11-28}$$

Then it is expected that the most likely response to an arbitrary (or noisy) input pattern $\underline{x}_a$ is:

$$r_j(\underline{x}_a) = M(t) \tag{11-29}$$

This is called the *degree of saturation* of the discriminator. Then, it is possible to define an *absolute confidence factor* CON(a), as the confidence in relation to the response to $\underline{x}_a$. Hence:

$$CON(a) = 1 - \frac{M(t)}{MAX[Z(\underline{x}_1)]} \tag{11-30}$$

As training increases M(t) approaches 2^N asymptotically, whereas MAX $[Z(\underline{x}_i)]$ approaches 1. Thus, CON(a) tends to 0 with training at a rate determined by $1/2^N$. Note that in tending for 0, CON(a) can temporarily assume negative values, depending on MAX $[Z(\underline{x}_i)]$. This again shows how the confidence in the system depends heavily on N. However, and this may be surprising, in practical cases with WISARD, for $R = 512 \times 512$, it was never necessary to have N greater than 8. Finally, it becomes of interest to estimate the rate of growth of M(t) with t by means of a highly approximate analysis.

Assume that a discriminator trained on a set of patterns $T = \{\underline{x}_1, \underline{x}_2, \underline{x}_3 \ldots \underline{x}_{t-1}\}$ is now tested with the tth pattern $\underline{x}_t$ giving a response $r(\underline{x}_t)$. This response is a measure of the N-tuples already set, hence if the discriminator is now trained on $\underline{x}_t$, the total additional number of bits set in the entire discriminator memory is $R[1 - r(\underline{x}_t)]$. Therefore, the increment in M(t), Δ M(t), is given by:

$$\Delta M(t) = \frac{1}{2^N} \cdot \frac{1}{K} \cdot R[1-r(\underline{x}_t)] \tag{11-31}$$

$$= \frac{1}{2^N} \cdot N[1-r(\underline{x}_t)]$$

since $K = \dfrac{R}{N}$.

From this it is possible to calculate:

$$M(t) = \frac{N}{2^N} \sum_{i=1}^{t} [1 - r(\underline{x}_i)] \tag{11-32}$$

To get a better approximate value for this relationship, using the same argument as in equations (11-25), (11-26) and (11-27) it is shown that:

$$\Delta M(t) \cong \frac{N}{2^N} [1 - (\frac{S_{st}}{R})^N] \tag{11-33}$$

where S_{st} is the similarity between the tth pattern and $\underline{x}_s$ where $\underline{x}_s$ is the pattern in T most similar to $\underline{x}_t$. The table below indicates some rates of saturation Δ M(t).

S_{st}/R	N=2	4	8	
0.9	.095	.086	.018	$\Delta M(t)$
0.5	.375	.234	.03	

Thus, again note the beneficial effect of increasing N as for a given degree of similarity S_{st} reduces the rate of saturation with increasing N.

Analysis of Responses to Shapes

This section refers to the discriminatory powers of a RAM single-layer system with respect to "shapes". These can be described as sets of patterns:

$A = \{ \underline{a}_1, \underline{a}_2, \underline{a}_3 \ldots \}$ for shape A

$B = \{ \underline{b}_1, \underline{b}_2, \underline{b}_3 \ldots \}$ for shape B

where $\underline{a}_j$ EA and $\underline{a}_k$ EA are patterns of the same "shape" translated and/or rotated to a new position. We wish to address the following problems:

1. position-free recognition;
2. coping with several shapes and occlusion;
3. finding the position and orientation of simple shapes.

To simplify matters, a particular example will be followed throughout this section and that is that there are two specific shapes related as shown in Figure 1.

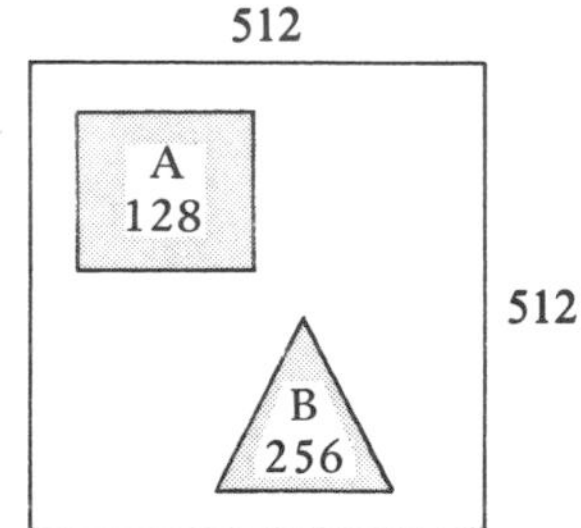

Figure 1. Examples of shapes

Thus, set A is the set of all 128×128 squares that could appear in a 512×512 frame and B a set of triangles with dimensions as shown.

The two shapes have the same area so that the possibility of mere area measurement is removed. Further, note that the black/white shading and the geometrically definable nature of the objects is arbitrary. The system is capable equally of dealing with textured objects which would defeat most geometrically orientated recognition techniques such as described in [13].

Position-free Recognition

For simplicity, it is assumed that the above two shapes can appear only in their "stable" positions with respect to the lower edge of the frame. This gives one orientation for square A and three for triangle B. The first step in selecting N is deciding on the size of the training set. This is guided by the fact that an arbitrarily placed object must provide an area of overlap with an element of its approximate training set which is greater than the maximum overlap between the two objects. A simple but tedious calculation shows that for this condition to hold the discriminators can be trained in a scan-like fashion with a total of 81 patterns for discriminator A (the square) while B requires 114 triangles with the long side horizontal and 242 with the short side horizontal.

N is selected to keep the system out of saturation. Using equation (11-32) and an approximation for $r_j(\underline{x}_i)$ as used in equation (11-33) but this time applied to equation (11-17), an equation for the rate of saturation $\Delta M(t)$, based on the overlap of the next training pattern and two others can be derived. These are, the one just above and the one just to the left, for the scan from left to right and top to below. Thus:

$$\Delta M(t) \simeq \frac{N}{2N} \left[1 - \left(\frac{S_1}{R}\right)^N - \left(\frac{S_2}{R}\right)^N + \left(\frac{S_{12}}{R}\right)^N\right] \tag{11-34}$$

where S_1 is the overlap area of the new training pattern with the one above; S_2 is the overlap area of the new training pattern with the one to the left; and S_{12} is the common area between S_1 and S_2 as before. This expression can be summed for the entire scan from a knowledge of the areas of overlap which need to be calculated for each stable position. Taking the triangle as the more stringent case, it is interesting to tabulate M(t), the percentage saturation as a function of N in a significant range:

N =	8	9	10	11
M(t) triangle =	> 100%	> 100%	80%	52%

This is a highly pessimistic calculation as the approximations themselves are pessimistic. However, the result indicates that $N = 10$ is the least value that can be chosen. The confidence with which the systems will distinguish between triangles and squares for $N = 10$ can be checked by using the worst overlap with its opposing class and the "worst fit" with its own class in equation (11-27). It turns out that the system responds with a minimum confidence level of 82% in response to a square and 81% in response to the most "difficult" triangle.

In practice, a scan-training technique (presented here because it is analytically tractable) would not be adopted necessarily. Indeed, start by selecting a feasible value for N and adjust the training so as to keep the system out of saturation, increasing N only if the eventual confidence levels fall near or even below zero (giving the odd erroneous classification). The training patterns would be determined by operator presentation of objects, which is particularly attractive in robotic situations where the operator, in general, communicates with the robot by means of examples.

Effect of Several Shapes and Occlusions

Much can be done by studying the responses of the system given constraints on combinations of patterns that can appear in the field of view. Here, the previous example is continued but the class of patterns that can be viewed is extended to the presence of any *two* objects in the field of view. Again, this is done for analytic simplicity, and initially it is assumed that no overlaps occur. Also, the response equation is simplified to take into account the two nearest training patterns as in the previous section. Equation (11-27) is the confidence factor equation. N is taken as 10. Using this it may be noted that the response of the "square" discriminator to a single square in the image, $r_s(\underline{x}_s)$, is:

$$82.4\% \leqslant r_s(\underline{x}_s) \leqslant 100\% \text{ (with a minimum of 82\% confidence)}$$

Similarly for the "triangle" discriminator and a single triangle, $r_t(\underline{x}_t)$:

$$81.8\% \leqslant r_t(\underline{x}_t) \leqslant 100\% \text{ (with a minimum of 81\% confidence)}$$

Now, if two objects are in the field of view simultaneously (a square and a triangle, non-overlapping) we have for this $\underline{x}_{st}$, say:

$$45.5\% \leqslant r_s(\underline{x}_t) \leqslant 54.2\% \text{ (with a maximum confidence of 18.9\%)}$$

and similarly for the triangle:

$$42.3\% \leqslant r_t(\underline{x}_{st}) \leqslant 54.2\% \text{ (with a maximum confidence of 19.27\%)}$$

The *range* of responses arises from the degree of overlap between the test pattern and the nearest training pattern. If the two objects were of the same kind, say both squares (ie $\underline{x}_{ss}$), these responses are modified as follows:

$$55.1\% \leqslant r_s(\underline{x}_{ss}) \leqslant 66.4\% \text{ (with a confidence range 21.8\% to 50\%)}$$

$$33.1\% \leqslant r_t(\underline{x}_{ss}) \leqslant 43.1\% \text{ (with the above confidence range)}$$

Very similar results are obtained if both objects are triangles ($\underline{x}_{tt}$), with $r_t(\underline{x}_{tt})$ being the highest.

Clearly, these results can be distinguished in two steps: first on the basis of the confidence level which identifies the situation group (one object, two similar objects, two dissimilar objects) and second the response itself which identifies the object(s) in the first two groups.

Considering now the question of occlusions, take the worst possible overlap between the triangle and the square, which is the case where the right angle of the triangle is coincident with one of the corners of the square. Predicted responses for the *two* discriminators turn out to be identical as the similarity areas with nearest training elements are the same. The response itself is, to ($\underline{x}'_t$), say:

$$r_s(\underline{x}'_{st}) = r_t(\underline{x}'_{st}) = 78.5\%$$

The predicted confidence level is *zero* and this fact alone, despite the much higher response, in comparison with $r(\underline{x}_{st})$, allows detection of the occlusion or quasi-occlusion as being a specific case of the presence of two objects.

Although in practice the number of discriminators might be increased, which then would be trained separately for multiple object classes and occlusions, the above results are indicative of the discrimination that is available without taking this step.

Position and Orientation Detection

In this section we merely point to the fact that the response of several discriminators, each trained on different positions or orientations of an object, may indicate with a reasonable degree of accuracy the actual position of the object. To take an almost trivial example, imagine a 16-discriminator 512×512-bit system trained on 128×128 squares, each discriminator having seen just the one square in one of the 16 possible disjoint positions. In use, with a square in an arbitrary position, the response of the discriminators will depend on the overlap area between an input square and those in the training set, allowing computation of the position of the square with some accuracy. This extends even to $N=1$ when a parallel mask-matching exercise is carried out. However, in order to carry out the position-finding exercise at the same time as retaining some discrimination between several objects, a higher valued N can be used.

With higher N values, the number of discriminators in the system can be reduced in the same way that each can be trained on a judiciously selected *half* of significant positions, allowing the system to provide a binary code for the actual position.

Clearly, all that has been said about position applies to orientation. On the whole, the system designer has merely to select a relatively low number of discriminators to pin-point key orientations and the value of N according to the principles discussed above and use the system as an interpolatory device to provide precision.

Alternative Structures for Future Development

So far in this system the structure that has been analysed is a "feed-forward" kind. That is, it consists of two sub-systems; the net itself and the decision machinery (which calculates MAX $[Z(\underline{x}_i)]$) where the former having received a "stimulus" image feeds data forward to the latter, which outputs a "response"; that is, the decision itself. Once the adaptation period is over this "stimulus-response" action is cominational; that is, the action of a trivial finite state machine with only one state. Much current work is aimed at the incorporation of the above system into a dynamic structure where feedback between the "response" and the "stimulus" is introduced. Part of the "response domain" thus becomes the "state space" of a finite state machine, and, if one views the image spaces as some sort of Turing machine tape, it becomes clear that the system extends its computational powers much beyond that of a single-layer net (where single-layer networks are static) and that the extent of this needs researching. Thus, these systems are "dynamic".

To be specific, work is progressing using the decision domain to control the size and position of the viewing window, as well as for recognition. This follows the work of Reeves [14] and Dawson [15], the latter having shown that letting the system control the zoom enables a scene of two-dimensional objects to be labelled in terms of size, position and orientation. However, in these cases position control was achieved either by a pre-programmed scan [15] or by small increments in horizontal and vertical positions [14]. In current work, the aim is to control the position of the window by *saccadic jumps*, which enables a system to take a series of "closer looks" before providing a final decision. Being a learning system, the saccadic jumps become context-dependent providing a powerful tool not only in object labelling in multi-object scenes, but also in the analysis of similarly shaped objects which differ in local detail.

A second line of enquiry into dynamic systems follows some earlier work on short-routed feedback [16], the design of adaptive language acceptors [17] and more recently, dichotomizers with increased sensitivity through feedback [18] and artificial perception in array-structured systems [19]. The common theme running through the above investigations is that a tendency towards stability in adaptive dynamic systems results in interesting processing properties. In current work this property is being harnessed through giving the decision domain an "alphabet" and translating the stable interaction with an input into linguistically structured statements. For example, the system may provide decision statements such as "The triangle is above the square" or "There are two triangles". This provides a direct, fast solution to some old scene analysis problems, and may provide the basis for the design of useful inductive learning engines for "expert systems" of the future. In passing, note that such feedback systems overcome "perceptron limitations" [2] and may contribute to an explanation for the success of living brains which, after all, have to get along on the basis of perceptron-like devices without the benefit of "fifth-generation" computer languages and structures.

References

1. McCulloch, W. S.; Pitts, W. H. A logical calculus of the ideas imminent in nervous activity. *Bulletin of Mathematical Biophysics* 1943, **5**, 15-33.
2. Minsky, M. L.; Papert, S. *Perceptrons: An Introduction to Computational Geometry* MIT Press, Cambridge, 1969.
3. Rosenblatt, F. *Principles of Neurodynamics* Spartan, Washington, 1962.
4. Aleksander, I.; Albrow, R. C. Pattern recognition with adaptive logic circuits. *Proceedings of IEE – NPL Conference on Pattern Recognition* 1968, **42**, 1-10.
5. Aleksander, I., Stonham, T. J., Wilkie, B. A. Computer vision systems for industry. *Digital Systems for Industrial Automation* 1982, **1** (4).
6. Widrow, B. Generalisation and information storage in networks of adaptive neurons. In *Self-organising Systems* eds Yovits *et al*; Spartan, New York, 1962.
7. Wilkie, B. A. *Design of a High-resolution Adaptive Pattern Recogniser* Thesis, Brunel University, in preparation.
8. *British Patent Application No. 8135939* November, 1981.
9. Bledsoe, W. W.; Browning, I. Pattern recognition and reading by machine. *Proceedings East. J.C.C.* 1959, 225-232.
10. Ullman, J. R. Experiments with the N-tuple method of pattern recognition. *IEEE Transactions on Computers* 1969, 1135.

11. Aleksander, I.; Albrow, R. C. Microcircuit learning nets: hamming distance behaviour. *Electronics Letters* 1970, **6**, 134-135.
12. Aleksander, I.; Stonham, T. J. A guide to pattern recognition using random access memories. *IEE Proceedings on Computers and Digital Techniques* 1979, **2**, 29-36.
13. Dessimoz, J. D., Kammennos, P. Software for robot vision. *Digital Systems for Industrial Automation* 1982, **1**, 143-160.
14. Reeves, A. P. *Tracking Experiments with an Adaptive Logic System* Thesis, University of Kent, 1973.
15. Dawson, C. *Simple Scene Analysis Using Digital Learning Nets* Thesis, University of Kent, 1975.
16. Fairhurst, M. C. *Some Aspects of Learning in Digital Adaptive Networks* Thesis, University of Kent, 1973.
17. Tolleyfield, A. J. *Some Properties of Sequential Learning Networks* Thesis, University of Kent, 1975.
18. Aleksander, I. Action-orientated learning networks. *Kybernetes* 1975, **4**, 39-44.
19. Wilson, M. J. D. Artificial perception in adaptive array. *IEEE Transactions on Systems, Man and Cybernetics* 1980, **10**, 25-32.

Chapter 12

Emergent Intelligence from Adaptive Processing Systems

I. ALEKSANDER

Department of Electrical Engineering and Electronics
Brunel University, England

Abstract: The n-tuple recognition net is seen as a building brick of a progression of network structures. The emergent "intelligent" properties of such systems are discussed. They include the amplification of confidence for the recognition of images that differ in small detail, a short-term memory of the last-seen image, sequence sensitivity, sequence acceptance and saccadic inspection as an aid in scene analysis.

Introduction

Artificial intelligence (AI) work in the 1970s has superseded earlier research into the mechanisms of intelligent behaviour based on pattern-recognizing neural nets. This largely originated in 1943 with the McCulloch and Pitts model [1] of the neuron. The main reasons for this demise are understood to be the following:

1. The "perceptron" limitations of neural nets were stressed by Minsky and Papert in 1969 [2]. These largely centred on the inability of some nets to carry out simple object-counting operations.
2. The writing of programs that have a surface behaviour which, if attributed to man, would be said to require intelligence, has become a well-developed and understood science [3].
3. The generality of the conventional computer structure discourages a study of architectures incompatible with it.
4. Neural nets are programmed through exposure to data and in this sense are "learning machines". It is argued that the age of the universe would be required to *learn* human-like intelligent behaviour and this is best avoided by more direct programming.

In this paper, it is argued that 1 is invalid as it applies only to a limited class of neural nets; 2 is one of many methodologies for studying mechanisms with an intelligent surface behaviour, another, the pattern recognition neural net, being proposed as having equal validity; 3 has been invalidated by the possibility of totally novel architectures being facilitated by advancing VLSI techniques whereby "intelligence" may be derived with greater cost-effectiveness than is offered by the conventional computer. The argument in 4 is appealing, but extreme. In the same way as algorithmic AI approaches provide a limited window view of intelligent behaviour, so can studies in learning neural networks. It is the knowledge-gaining efficacy of the research that is at stake. It is argued in this paper that this may be greater through a study of neural nets and that, indeed, the vistas over mechanisms of intelligent behaviour may be wider. The central argument evolves around the concept of "emergent intelligent properties" of a system, as will be discussed in the next section.

Emergent Intelligent Properties

The question of emergent properties of machines and their relationship to parts of these machines has recently been lucidly discussed by Gregory in 1981 [4] . The architecture of the brain, that is the parts of the brain machine and their interconnections, is undeniably based on building blocks that consist of neural tissue. Intelligent behaviour is one *emergent property* of such tissue.

In AI studies any intelligent behaviour exhibited by the computer is *not* an *emergent property* of the computer itself, but rather, of the program. The latter however needs to be argued – it is not as obvious as one would like it to be.

On the one hand, it could be argued that programming a computer is like painting a canvas. No one would believe that the image on the canvas is an *emergent property* of the canvas and the paints. It is merely an expression of the artist's intent. The fact that a computer pronounces "the black box is in front of and slightly to the left of the green pyramid", the canvas-and-oil school would argue, is merely a confirmation of the programmer's (artist's) original intention.

On the other hand, the AI fraternity would argue that the statement issued by the machine is an emergent property of the *rules* obeyed by the program, rules which the programmer has chosen with great care. Are these rules just the spots of a pointillist painter or the true atoms of a system whose interaction creates an effect not contained in the atoms themselves?

This paper does not seek to resolve this question. It simply presents alternative technical structures which are both technologically interesting, because of their advanced performance, and fundamentally attractive as intelligent behaviour is less dependent on a highly directed choice of pre-programmed rules.

But what constitutes this "fundamental attraction"? The alternative structures are indeed neural-like while at the same time being technologically feasible and properly adapted to implementation. However, the fundamental attraction comes from the fact that the emergent properties of these structures may be more general than those ensuing from a carefully chosen set of rules. For example, the emergent properties of such a system may be the ability to append proper linguistic labels to *all* the examples shown in Figure 1.

Figure 1 Diversity in Vision Work

Even to the most enthusiastic AI practitioner this appears a difficult task because of the exploding multiplicity of rules that may be required. Clearly, this paper will not present a structure that meets this specification either. It will, however, examine to what extent artificial neural net structures are likely to have emergent properties that head in the stated direction without an exploding set of elements.

Level 0 Structures: Image Classifiers

It will be shown that increased prowess in the emergent properties results from a progression of physically distinct structures, each level of complexity relying on an understanding of the previous level.

At the lowest level (0) the nature of the simple artificial neuron is defined (much of this being available in the published literature) as is the nature of the simplest network structure: the single-layer net as shown in Figure 2. This, indeed, is the structure of the WISARD system [5] currently used for high-performance image identification. The single-layer net is described here in terms convenient for the rest of the paper. A full technological and theoretical description of the principles involved may be found in the literature cited.

The artificial neuron is defined as a bit-organized Random Access Memory (RAM) where the address terminals are the synaptic inputs (N in number), the data output terminal is the axonic output, whereas the data input terminal is the "dominant synapse", as in the terminology of Brindley [6] . During a training phase the RAM stores the response (0, 1) to the binary N-tuple at the synapses as required by the dominant synapse. During a "usage" period the last stored response for each occurring N-tuple is output at the axon. Clearly, variants on this are possible: the neuron can be made to store and output a number, if word rather than bit organized. Also a statistical function could be used to control the output, as in Uttley [7] . However, these schemes may be shown to be variants of the "canonical" neuron discussed above.

A single-layer net consists of C sets of K artificial neurons each. Each set is called a *discriminator*. Each neuron in a discriminator receives the *same* data input bit. Thus, each discriminator has KN synaptic inputs assumed to receive the input image. The connection of the bits of an image to this vector is one-to-one, but arbitrary. Once such an arbitrary connection has been determined it remains fixed

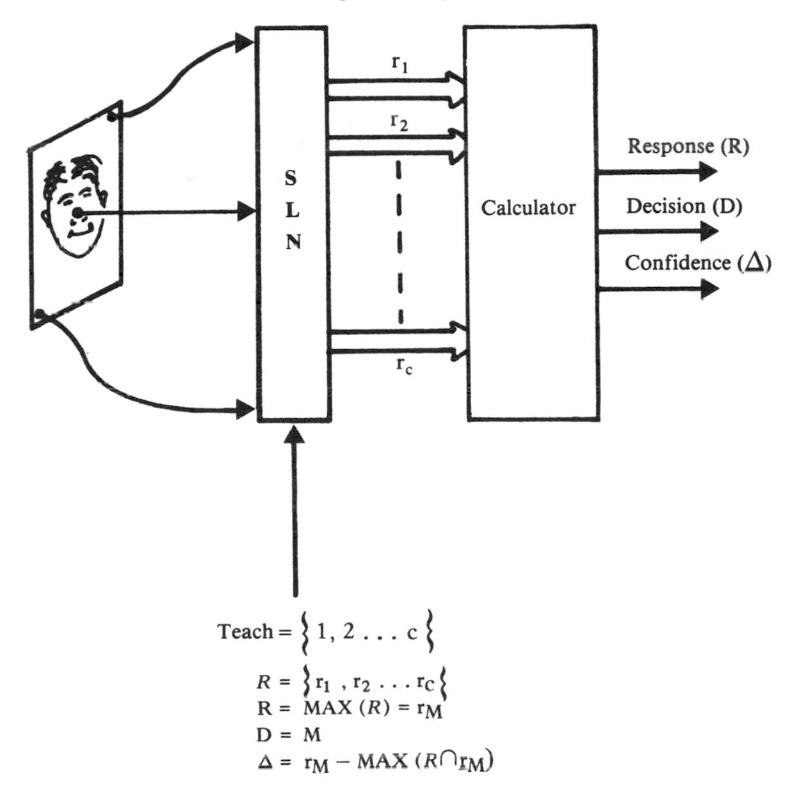

Figure 2 Level 0 Structure

for the net. The same connection may be used for all discriminators (coherent net) or not (incoherent net). Such coherence may be shown to have little effect on behaviour.

Training involves one discriminator per class to be distinguished. For a representative set of images a logical 1 is fed to all the elements of the discriminator corresponding to the known class of the current image. When an unknown pattern is present at the input, the number of 1s at each discriminator output (out of a total of C) is counted. The *first properties,* proven and discussed in Aleksander [8], should now be stated for completeness.

Property 0.1: Recognition and Confidence Level

The system transforms an input binary vector $\underline{I}$ of magnitude KN bits into an output vector of C numbers (responses), each such number being the number of 1s at the discriminator output, expressed as a proportion of K. The discriminator with the highest response is said to classify the image. The classification has been shown to be predictable through a somewhat awkward set of expressions that can be simplified to the following. The response of the jth discriminator r_j is most likely to be:

$$r_j \simeq \left(\frac{S_j}{KN}\right)^N \tag{12-1}$$

where S_j is the overlap area between the unknown incoming vector and the *closest* pattern in the training set for the jth discriminator. The fact that this is *not* a mere pattern match and that performance depends on N is demonstrated by the *confidence* with which a decision is made. Confidence is defined as:

$$\text{CON} \simeq \frac{r_1 - r_2}{r_1} \tag{12-2}$$

where r_1 is the strongest discriminator response and r_2 that of the next one down. This becomes:

$$\text{CON} \simeq 1 - \left(\frac{S_2}{S_1}\right)^N \tag{12-3}$$

where S_2 is the overlap of the unknown with the closest element in the second most strongly responding discriminator, and S_1 with the first. The confidence (and, as argued in Aleksander [8] , avoidance of saturation of the system) increases with N. Against this it must be realized that the cost of the system grows exponentially with N. Two further particular emergent properties should be noted.

Property 0.2: Acceptance of Diversity

The training set for a particular discriminator can be highly diverse (examples may be found in papers already quoted: many orientations and expressions of a face, rotations and positions of a piece part). The system performs its classification *without a search*. The response vector may be generated after one pass through the storage system. This property is most rare in AI work in general.

Property 0.3: Sensitivity to Small Differences

The system can *both* differentiate or group together images with small differences between them. For example, experiments on WISARD have demonstrated that a distinction may be obtained for smiling and unsmiling faces or, if required, faces may be recognized independently of their expression. This is achieved merely by task definition through training, where the system accommodates the trainer's requirements as effectively as possible. It would be most difficult to find a set of rules *a priori* which can differentiate between any two faces, which may subsequently be modified easily to react differently to different expressions. Finally, it should be noted that a single-layer network can distinguish the presence or absence of the dog in a "Marr-Dalmatian scene" probably with greater ease than man, as its action can be forcefully directed towards the task. The important fact is not that the system recognizes the dog but, that it is *sensitive* to such complex data.

More interesting, perhaps, is a speculation on whether a SLN might be able to recognize the face of the politician in Figure 1. The transfer from a grey-tone image to an outline is, apparently, not a learnt function. The classical approach to explaining this transfer phenomenon relies on the supposition of the existence of a neural system in the visual pathway that performs a differentiation operation. This would provide the similar output for both representations. Neural nets that are pre-programmed to perform the differentiation are well understood, and will not be evoked in the progression of structures presented here.

Level 1 Structures: Decision Feedback

The added ingredient of all structures that proceed is the inclusion of feedback between the response vector and the input, as shown in Figure 3. In this case, the output vector bits are mixed with the input vector to create a new input vector. The latter is then distributed among the N-tuple inputs as before. This technique was first reported by Aleksander in 1975 [9], and has more recently been investigated further using the WISARD system and the following method. Figure 4 shows the experimental arrangements. Two cameras are used, one for input and the other for a display of the discriminator responses. These two images are mixed to form the input

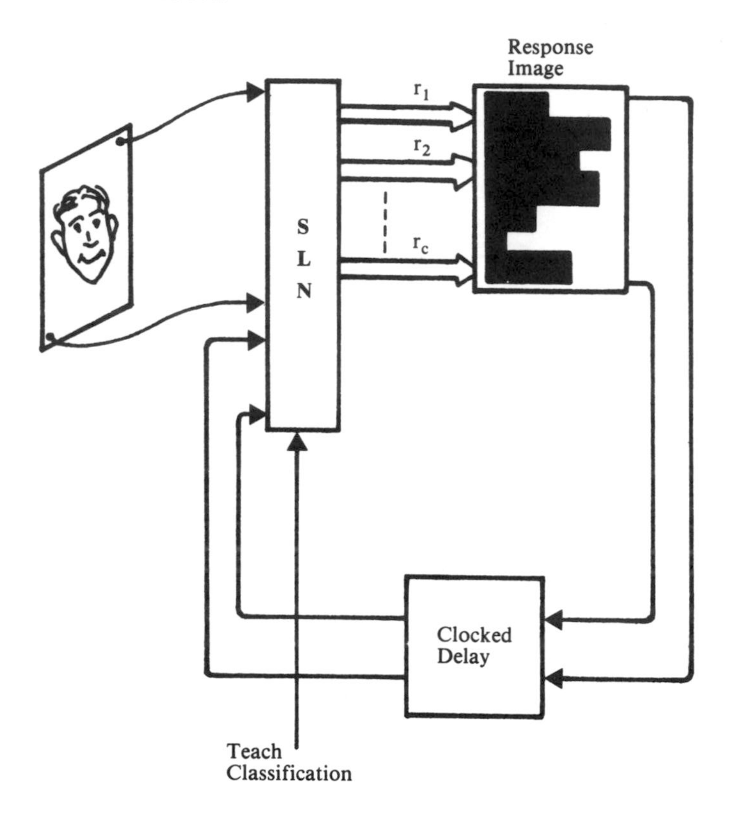

Figure 3 Level 1 Structure

field to WISARD. A feedback clock is included in the system allowing step-by-step recordings. This connection leads to the next emergent property.

Property 1.1: The Amplification of Confidence

This is particularly useful when discriminating very small differences between images. For example, the scheme can distinguish with high confidence (after a few steps of feedback) whether a small speck is on the left or the right of the image. Clearly a single-layer net would have a limited confidence since $S_1$and S_2 in equation (12-3) would be very close in this situation. The results obtained with WISARD are shown in Figure 5 where a speck consisting of 5% of the image is sought either on the right or the left of the image

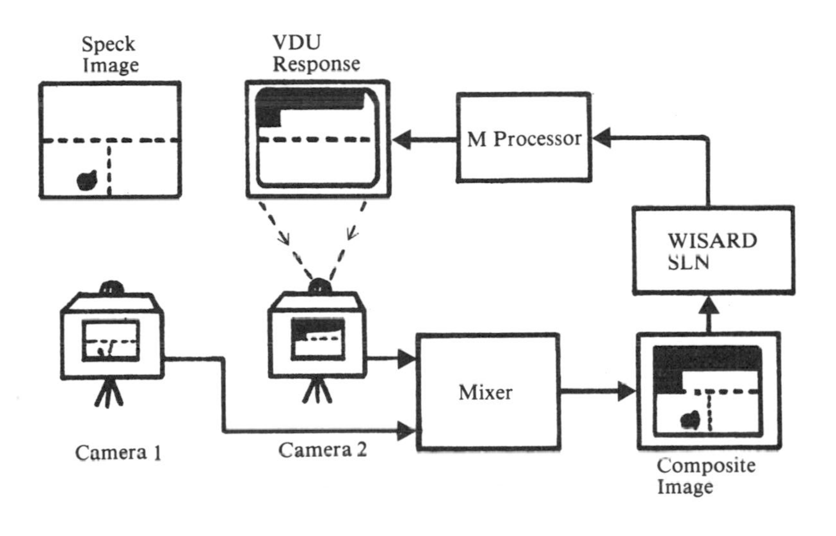

Figure 4 Level 1 Experiment

(where N = 6). The maximum confidence without feedback would be about 30%, whereas the feedback system reaches over 90%.

Property 1.2: Short-term Memory of Last Response

The results also clearly show a second emergent property common to many feedback structures: a short-term memory. This relates to the last-made positive decision which persists even when the image is removed. In the context of AI, the only importance of this structure is that it overcomes one of the perceptron limitations brought to light by Minsky and Papert [2] . It is clear that a neural perceptron structure with no feedback would not be a suitable speck detector in the way discussed above. The discontinuation of work on neural systems in the 1970s was due to the discovery of several limitations of this kind, which, as partly shown in this paper, only apply to some (mainly level 0) artificial neural structures.

Level 2 Structures: a Preamble

One property noticeably lacking in the structures described so far is the ability to regenerate or recreate learnt patterns. A structure

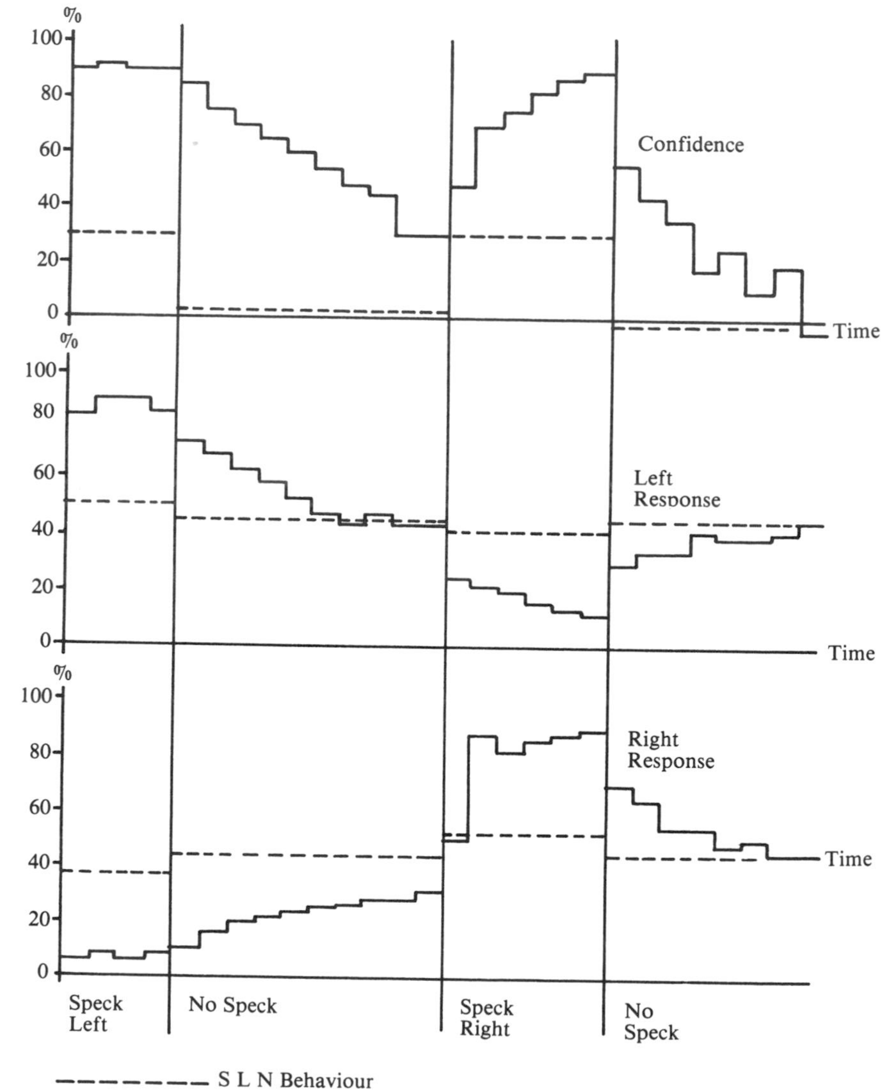

Figure 5 Confidence Amplification

whose emergent property is that, at any point in time it has an inner *state image* in the following. Let us suppose that the response vectors of a single-layer net are no longer trained on desired 1-in-C classification vectors, but that the training consists of applying another *image* vector to the C-training terminals. The emergent property of a system of this kind is that it could respond with prototypes of patterns (having been trained on them) to variants of input images. For example, the input could be a square and a triangle in any position in the field of view while the output could be the prototype of each of the two classes. A typical result of this form of training is shown in Figure 6.

It has been shown theoretically in Aleksander [8] that a 2-discriminator SLN operating on a 512×512 image with $N = 10$ and $C = 2$ may be expected to recognize squares and triangles both having an area of 16,384 pixels as shown appearing in any "rest" position. Experiments have supported this, but shown that the recognition takes place with a low degree of confidence, in some cases. In practice, the dotted areas in Figure 6 would have about 80% intensity with the black areas being 100% dense.

Note that in the proposed structure, the output can produce K grey levels at each pixel, which may be excessive. In fact, it is possible to reduce the size of the net to a binary output pixel, in which case each output cell only "sees" one N-tuple of the input image. This can lead to interference by the last-seen pattern. It has been shown that if the property of "forgetting" is added to the net (note that this is not an emergence property, but a change in structure), this effect can be compensated.

Property 2.1: Short-term Image Memory

So far, we have only pointed to lemmas of a level 2 structure. The final item is the introduction of feedback between the output and the input as for level 1 structures. Clearly, this has the property of increasing confidence, for instance increasing contrast in the output image. This, incidentally, now becomes the *state image*, so that the short-term property of the level 1 net translates into a *short-term memory* of the prototype of the last-seen image which fades with time, as shown in Figure 6.

Finally, a variant can be added to the level 2 structure in the way that the dominant synapse or teach terminals are connected. This

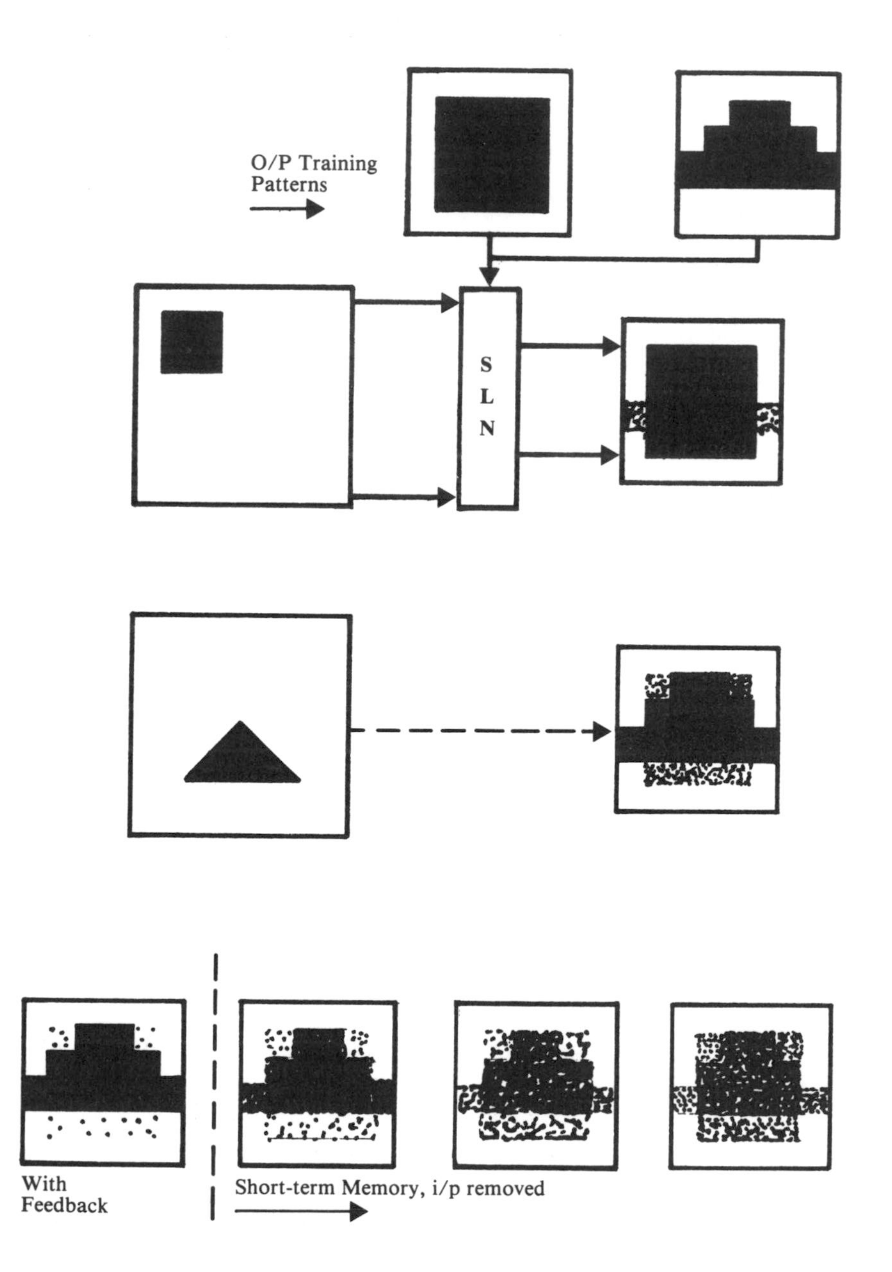

Figure 6 Level 1/2

technique simply feeds image information from the *input* to the teach terminals, as shown in Figure 7. This circumvents the rather unnatural situation of needing a separate channel for training.

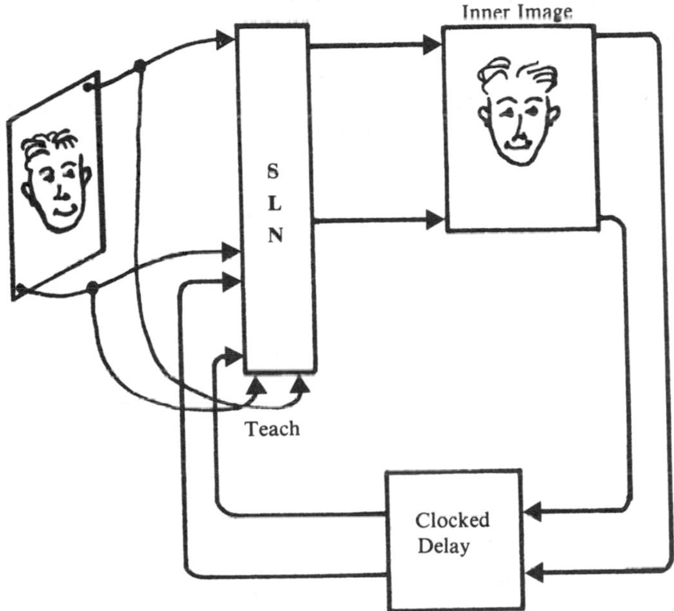

Figure 7 Level 2 Structure

Level 2 Structures: Emergent Properties

Property 2.2: Artificial Perception

This property has been described by Wilson in [10], albeit in respect to an array network, where specific feedback connections (from a nearest-neighbour arrangement) were used. The essence of this is that, starting from random states, the system rapidly enters state replicas of images seen during training. This link occurs *only* for learnt images, non-learnt ones lead to random state changes. This emergent property is thought to be the key to "modelling of the environment" within neural nets.

Property 2.2: Sequence Sensitivity

The dynamic property relates to the *chaining* ability of the system, which enables training the system to form associations between pairs of input images or longer sequences. The training goes as follows. Start with, say, a blank state b, and input i_1. This, upon applying a training pulse, transfers i_1 to the state. The input is then changed to i_2 and so on. The end of the sequence again can be identified with a blank. This training sequence can be summarized as follows:

Step number	*Input*	*State*	*Next state*
1	i_1	b	i_1
2	i_2	i_1	i_2
3	i_3	i_2	i_3
.	.	.	.
.	.	.	.
n	b	i_{n-1}	b

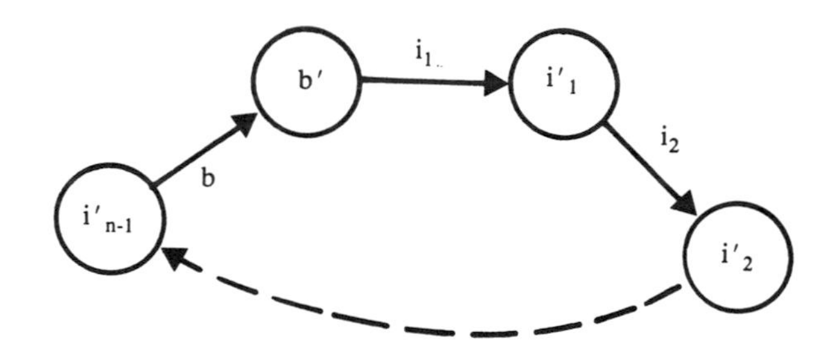

Figure 8 Sequence Following

This results in a finite state machine, approximately as shown in Figure 8. The approximation comes from the interference effect. Experiments with simulated systems show that a version of the input sequence, with degradation due to interference occurs if the same input sequence is presented a number of times.

Property 2.3: Lexical Acceptor

The chaining property may be extended further. Assume that the two training sequences are * J O H N * and * L O N D O N * (* is another, more pictorial, representation of a blank). The resulting state diagram is as shown in Figure 9. It should be stressed that such

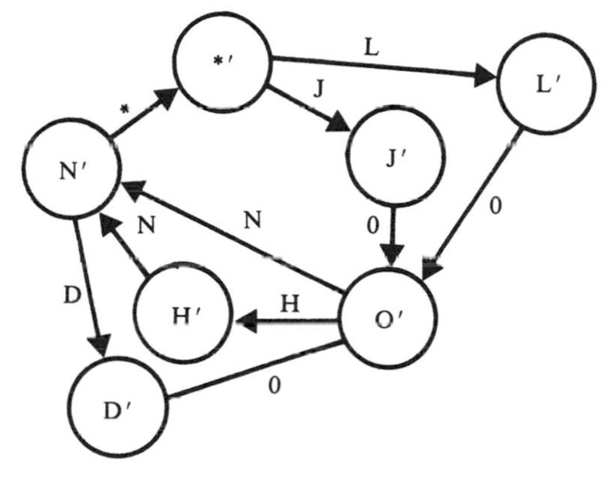

Figure 9 Lexical Acceptor

diagrams are incomplete and the departure from the taught sequences remains unspecified. Acceptance in terms of this kind is seen as a state sequence that is sympathetic to the input sequence. Clearly, a departure from the stored sequence indicates non-acceptance.

Property 2.4: Image-name Associations

If the input image was split so that an image part (constant) and a set of name symbols (time varying) were applied simultaneously to be input of the net, cross retrieval between the image and its corresponding name may be achieved. Let IM_i be the (say) facial image of *JANE* and IM_2 be the facial image of *JON* and the net trained on the two as before, then the state diagram in Figure 10 results. The emergence property is clearly the concept that if IM_1 is presented to

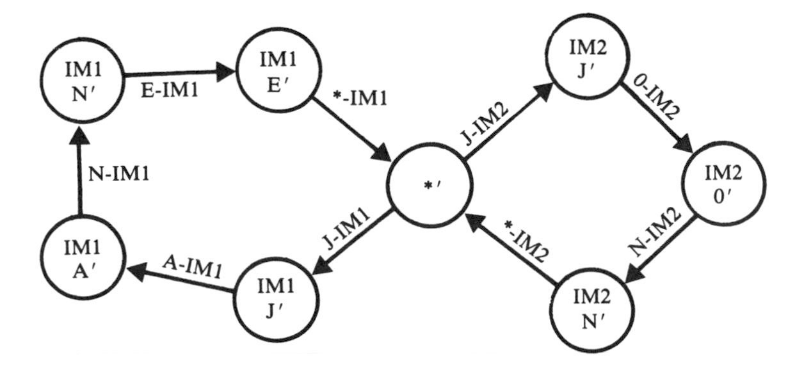

Figure 10 Image/Sequence Relationship

the system, it will enter the JANE cycle, whereas for IM_2 it will enter the JON cycle. Also given a repeated presentation of JANE, IM_1 will be recreated as will IM_2 for JON.

Level 3 Structures: the Attention Mechanism

This is shown in Figure 11. The SLN has two specific sets of outputs: the first deals with a decision based on the incoming pattern in the usual way, whereas the other feeds back to the input frame. The input frame itself is somewhat special: it contains an area of high resolution embedded in a field of lower resolution. The high-resolution area can be moved around through instructions fed to a window position-controller as shown.

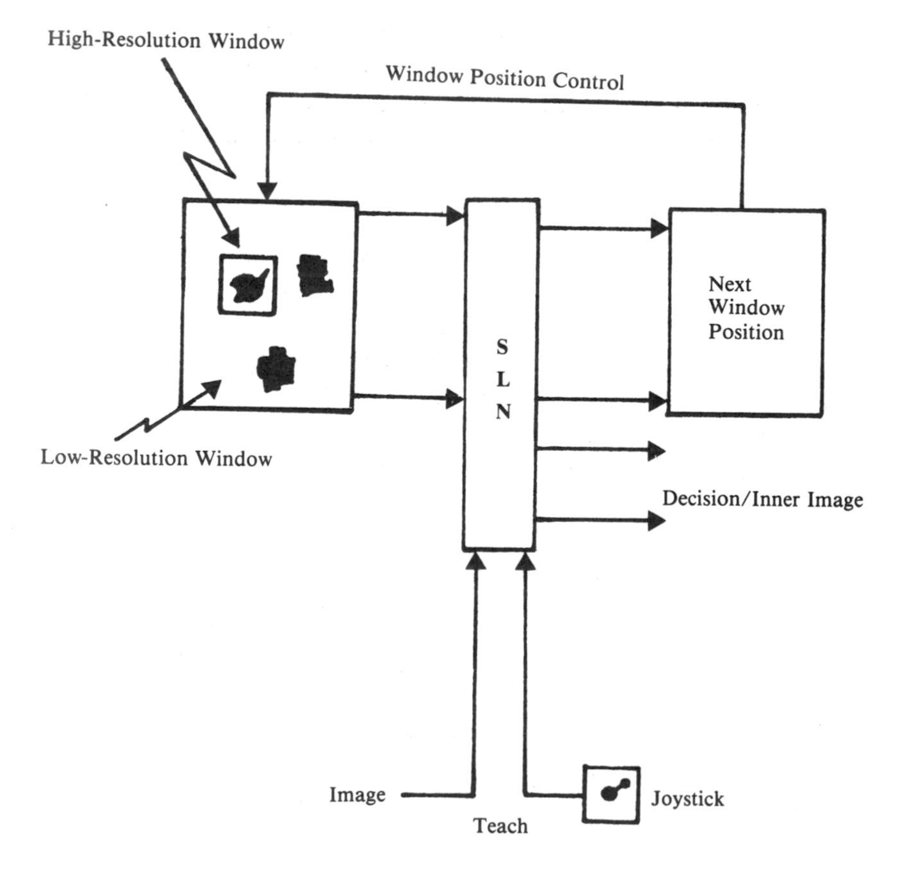

Figure 11 Level 3 Structure

Emergent Property 3.1: Blob Finding

A joystick input can be used to train the central window to move in the direction of a blob and stand still only when the latter is property centred. This is a more efficient way of obtaining position indepen dent recognition than the scan-training procedure discussed by Aleksander [8] . All that is needed is that the decision output be trained once the blob has been found.

Emergent Property 3.2: Saccadic Jumps

Should there be several blobs in the field of view, it is clearly possible to teach the system to centre on these one at a time, in some taught order, for example, clockwise or in a scanning motion. A study of level 3 structures has received only scanty research attention in the past [11] where it was shown that scenes consisting of several blob-like structures could be properly described as a level 3 system combined with conventional computing power. A deeper study is the subject of a current investigation. The aim of such study is to investigate an enhanced level 3 structure in which the SLN is operating as part of both a level 2 loop and a level 3 loop. It is possible to speculate on some emergent properties of such systems.

Speculative Emergent Property 3.3: Feature Sequence Detection

The level 2/3 system may be taught to take saccadic jumps from feature to feature in an image. The level 2 part would be able to interpret this sequence as belonging to a single object or scene. For example, the system may be taught to label a sequence of saccadic jumps between the corners of any triangle as "triangle", and, through the same process, distinguish it from a square.

Speculative Emergent Property 3.4: Disambiguation

The same scheme can be used to centre the high-resolution window on features that determine, say, whether one object is in front of another. This property may also unscramble some of the 2½ D letter heaps shown in Figure 1.

Conclusion

Clearly, the progressive series described does not end at level 3.

Interesting possibilities emerge as ever deeper nesting of the structures described takes place, and it is the objective of much current work to bring these to light.

In this final section it may be worth drawing attention to some important parallels, or lack thereof, between the emergent properties of artificial neural nets and living systems. These are as follows:

1. The parallels begin to become interesting when feedback becomes part of the structure. The precise training of level 0 nets appears artificial, and it is only at level 2 that particular exposures of the system to data through a single channel leads to associations of the type found in living systems.
2. Interesting feedback loops are being discovered by physiologists that appear to have some of the characteristics of level 2/3 structures. For example, loops controlling the eye muscles are well known and reminiscent of level 3 structures, whereas Barlow in 1982 [12] has reported recently that feedback between area VI of the visual cortex and the lateral geniculate body is under greater scrutiny. This is reminiscent of level 2 structures and may have a parallel role.
3. The Marr-Dalmatian experiment clearly points to an interaction between the immediate discriminant (low level) process of the human visual system and a higher level of processing. In terms of the structures discussed here this would be equivalent in level 2 to an inability, at first, to find the appropriate classification (or feedback cycle). Movement, or some knowledge of the presence of the dog, is equivalent to an aided entry into the recognition cycle which, once entered, is easily entered again.

Clearly, much work remains to be done with structures at level 3, composite structures and beyond. However, referring to Figure 1, it is felt that the pattern recognition neural net approach provides an interesting alternative to the AI paradigm, where higher level structures may have a range of emergent intelligent properties that are wider than any single AI programming methodology.

References

1. McCulloch, W. S.; Pitts, W. H. A logical calculus of the ideas imminent in nervous activity. *Bulletin of Mathematical Biophysics* 1943, **5**, 115-133.
2. Minsky, M. L.; Papert S. *Perceptrons: An Introduction to Computational Geometry* MIT Press, Cambridge, 1969.

3. Boden, M. *Artificial Intelligence and Natural Man* Harvester Press, Hassocks, 1977.
4. Gregory, R. *The Mind in Science* Weidenfeld & Nicholson, London, 1981.
5. Aleksander, I.; Stonham, T. J.; Wilkie, R.A. Computer vision systems for industry. *Digital Systems for Industrial Automation* 1982, **1** (4), 305-320.
6. Brindley, G. S. The classification of modifiable synapses and their use in models for conditioning. *Proceedings of the Royal Society of London, Series B* 1967, **168**, 361-376.
7. Uttley, A. M. Conditional computing in a nervous system. *Mechanisation of Thought Processes* London, 1959.
8. Aleksander, I. Memory networks for practical vision systems. *Proceedings of the Rank Prize Fund International Symposium on the Physical and Biological Processing of Images* London, 1982.
9. Aleksander, I. Action-orientated learning networks. *Kybernetes* 1975, **4**, 39-44.
10. Wilson, K. J. D. Artificial perception in adaptive arrays. *IEEE Translations on Systems, Man and Cybernetics* 1980, **10**, 25-32.
11. Dawson, C. *Simple Scene Analysis Using Digital Learning Nets,* Thesis, University of Kent, 1975.
12. Barlow, H. B. Understanding natural vision. *Proceedings of the Rank Prize Fund International Symposium on the Physical and Biological Processing of Images* London, 1982.